GW00691628

Air-Handling Systems Ready Reference Manual

Air-Handling Systems Ready Reference Manual

David L. Grumman, P.E.

McGRAW-HILL BOOK COMPANY

New York   St. Louis   San Francisco   Auckland
Bogotá   Hamburg   Johannesburg   London   Madrid
Mexico   Montreal   New Delhi   Panama   Paris
São Paulo   Singapore   Sydney   Tokyo   Toronto

Library of Congress Cataloging-in-Publication Data

Grumman, David L.
  Air-handling systems ready reference manual.

  Bibliography: p.
  Includes index.
    1. Ventilation—Equipment and supplies. 2. Air
conditioning—Equipment and supplies. 3. Heating—
Equipment and supplies. I. Title.
TH7681.G78 1986   697   86–34
ISBN 0-07-025072-3

1234567890 KGP KGP 8932109876

ISBN 0-07-025072-3

*The editors for this book were Joan Zseleczky and Barbara B. Toniolo,
the designer was Naomi Auerbach, and the production supervisor
was Thomas G. Kowalczyk. It was set in Primer by Techna Type.*

Printed and bound by Kingsport Press.

Contents

Preface

This manual is about air-handling systems and equipment, and it includes some simple techniques and measures for improving air-handling system performance. It is directed primarily toward systems commonly found in large commercial and institutional buildings such as hospitals, schools, colleges, libraries, office buildings, and stores. While the manual is intended for the less experienced hands-on operator of such systems, more experienced operators and supervisory personnel will also find it useful as a reference or a training tool.

An *air-handling system* is an assembly of components that circulates and treats air. Treatment of air may include filtering or cleaning, heating, cooling, humidifying, dehumidifying, speeding up, slowing down, pressurizing, and diffusing. The purpose of all of these is to help achieve a safe, healthful, and comfortable indoor environment. This manual will help you continue to meet that purpose in an energy conscious way.

Air-handling systems and equipment account for more energy use in buildings than is generally realized. This is especially true of the larger types of air-handling systems with many control zones. A difference of a degree or two in the setting of an airstream temperature may not seem like a great deal, but in actual fact it can make a considerable difference in how much energy an air-handling system and its components use. Therefore, it is important for the operator to understand the basic elements of air-handling systems and to recognize conditions that suggest energy-saving modifications to them.

This manual starts right off by describing a few simple operating and maintenance procedures (O&Ms) that can be put into practice to yield an immediate improvement in energy efficiency. Chapter 2 outlines and describes the components of air-handling systems. Chapter 3 enumerates the characteristics, advantages, and disadvantages of system types commonly in use. There is a separate chapter on energy conservation measures (ECMs)—which are more sophisticated than O&Ms. These are of moderate cost to implement, but they can still, under the right conditions, be worthwhile since they yield a favorable payback. The final chapter is on instrumentation: typical devices used with air-handling systems, how they're used, who makes them, and approximately what they cost. The book concludes with a bibliography for further study on the subject.

The list of O&Ms and ECMs outlined in this manual is not intended to be all-inclusive. However, the O&Ms and ECMs presented here were selected based on evidence that they are among the most common and effective of the procedures and measures that have been practiced in large buildings around the country since the energy crisis began.

The author would like to acknowledge the help received from the following individuals in preparing this book: Daniel L. Doyle, for his detailed work in developing the instrumentation section; Alexander S. Butkus, Jr., for advice on the book's concept and on technical matters; Vilma Barr, for guidance on the publishing process; Kathryn Teskoski, for preparing some of the illustrative artwork; and Virginia Skeele for typing and assembly.

Air-Handling Systems
Ready Reference
Manual

Operating and Maintenance Procedures to Improve Energy Efficiency

Operating and maintenance (O&M) procedures are energy efficiency improvement steps that can be taken by in-house staff. By definition, they cost little or nothing to implement. Often, O&Ms simply involve improved operating schedules and settings or better maintenance techniques.

The O&Ms mentioned below are relatively simple manual measures; they were selected as the ones most likely to yield significant improvements in energy efficiency. By no means should this list be regarded as a complete one, however.

O&Ms VERSUS ECMs

It will be noted that some of these O&Ms are similar to the more-automated energy conservation measures (ECMs) outlined in a subsequent section. The fact that these O&Ms *are* manual procedures is the main reason for their low cost; automating them, though increasing their effectiveness as energy savers, also increases their capital cost. They are then classified as ECMs rather than O&M procedures.

RECOMMENDED O&M PROCEDURES

O&M Procedure 1: Manual Start/Stop

What it is   The first procedure is one of the simplest and most obvious—but it should not be overlooked. Operators should con-

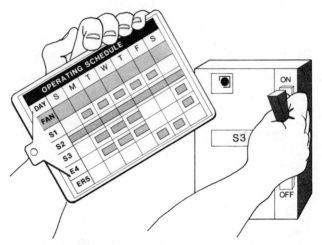

FIG. 1-1   Air-handling equipment left running unnecessarily is a frequent cause of energy waste.

sciously review the present procedures followed in their facilities to see if all is being done to *minimize energy use by limiting the amount of time equipment operates.*

Examples   All we're talking about here is shutting down equipment when it does not have to be on. This includes not only main air-handling system fans but also the many exhaust systems that are not tied into (interlocked with) the main systems. Toilet exhausts, for example, are often controlled by an independent switch or starter that must be manually flipped to start or stop the device. If toilets are not utilized during certain hours, are the fans exhausting them shut down during those hours? What about storage room exhaust fans that may be kept running, or even the hood exhausts of kitchen ranges? Who is responsible for shutting them down when the kitchen is not in use? Does the cook really care?

The same philosophy could also be applied to other equipment related to the air-handling system that may not be under automatic control—pumps, for example. Do coil water heating pumps really need to be kept running when the fan shuts down—or when the outside temperature rises above 45 or 50 °F?

Action to take   If it hasn't been done already, someone needs to sit down and list each piece of air-handling or related equipment in the building and its required hours of operation. Take the example of a fan; data listed should include fan tag number (or some other identification, like *supply fan*), area served, and occupancy hours for the various day types (weekdays, weekends, and/or holidays). The existing fan operating schedule should also be entered. Then the recommended start and stop times for the fan should be listed, taking into account space comfort levels and individual idiosyncrasies. A suitable form, with sample entries, is shown as Table 1-1.

TABLE 1-1 Schedule of Modified Air-Handling System Operation

| SYSTEM NUMBER | AREA SERVED | OCCUPIED TIMES | | NUMBER OF PEOPLE | OPERATING HOURS | | COMMENTS |
		HOURS	DAY TYPE		EXISTING	PROPOSED	
S-2	Administration	8 A.M.–5 P.M. / 9 A.M.–1 P.M.	Mon–Fri / Sat	30 / 10	5 A.M.–12 P.M. / 5 A.M.–4 P.M.	7 A.M.–5 P.M. / 8 A.M.–1 P.M.	
S-5	Operating room suite	7 A.M.–3 P.M.	Mon–Fri	35	24 hr/day Mon–Sun	6 A.M.–4 P.M.	

Then one needs to assign responsibility to specific individuals, by shift, for seeing that this equipment is indeed started and stopped on the prescribed schedule. An explanation of why the procedure is being implemented should be given as well. One then needs to check (spot checks are usually sufficient) on whether this procedure is actually being carried out.

SUMMARY OF ACTION STEPS

1. Identify equipment and list run times (Table 1-1)
   a. Present
   b. Proposed (anticipate comfort or people problems)
2. Inform operators of new schedules (and tell them why)
3. Check to see if followed

O&M Procedure 2: Fix Leaky Dampers

What it is *Dampers that leak air when they're supposed to be closed cause energy waste.* This can occur not only with outside-air dampers but also with return-air and mixing dampers. This section describes these situations, how they can be spotted, and what action to take.

Examples

Outside-air damper   The most obvious energy waster is the poorly fitting outside-air damper. Any opening provides a path for outside air to enter the building. Although the air-handling components are immediately adjacent to the outside damper, wind or, in taller buildings, "stack effect" can force this leaked air past the fan system itself. (In addition to energy losses, there is a danger that freeze-up could damage some air-handling components.)

If the fan system runs with the outside and exhaust dampers shut and those dampers leak, you may be drawing in outside air when there is no need to do so. It wastes energy and costs money to heat this air. (A fan system might be run in this mode for periodic nighttime heating or for morning warm-up.)

Return-air damper   It isn't just the outside damper that causes energy waste when it leaks, however. The return-air (or recirculated-air) damper can also cause waste if it leaks when it's supposed to be shut during mild weather, when the fan system is operating on 100 percent outside air. This might occur during mild weather (that is, 55 °F to 65 °F) when the mixed-air temperature (MAT) controller is attempting to maintain the airstream at 55 °F for cooling purposes. In such cases, the MAT becomes warmer than it needs to be, and mechanical cooling (that is, central chiller) will be needed sooner than it would otherwise.

For example, if it is 55 °F outside and, because of return-air damper leakage, a 60 °F MAT results, the building receives inadequate cooling and a chiller must be turned on at a lower outside temperature than would otherwise be necessary.

Mixing dampers   Another example is that mixing dampers in the hot and cold airstreams of air-handling systems (such as those on multizone units) can also affect energy consumption when they leak. A leak of 10 or 20 percent of total air flow through a hot-deck damper that's supposed to be closed into a cold airstream results in a mixture less able to cool. A portion of the energy expended in the cooling and/or heating coils is thus wasted.

How to spot leaky dampers

Observation   The first and most obvious way to locate leaky dampers is by inspection. Go into the air-handling system plenum with the outside dampers shut and take a look: Can you see spaces or daylight between the damper edges or around the ends? During cold weather, can you feel cold air coming through the wall opening the damper is supposed to close off? Try also to observe the situation when the fan is running with the outside-air damper closed. Inspection will give you a general idea of the magnitude of the problem you have.

Measurement and calculation   You can attempt to measure air flow directly, but this is difficult with low air flows. A more precise method of determining leakage is by measuring temperatures and using temperature differentials to figure out how much leakage there is. This is best done when the temperature spread is as large as possible—that is, when it is cold outside. This technique can be applied not only to outside-air leakage but to return-air and hot-deck and cold-deck damper leakage as well.

How to measure damper leakage   For instance, when a fan is operating on 100 percent recirculation, measure the outside-air temperature (OAT), the return-air temperature (RAT), and the mixed-air temperature (MAT). Be sure the readings are accurate, especially the MAT. (It is not advisable to use the thermometers

permanently mounted in the airstream unless you *know* they're accurate and/or have been calibrated.) Take care to ensure that you measure a representative average of the MAT; you need to do this because stratification (uneven mixing of airstreams) can occur. If you find the average MAT to be, say, 7 °F less than the RAT, you know you've got some outside-air (OA) leakage. How much? The following example shows how you can determine that.

Say your temperature readings are:

$$RAT = 76 \text{ °F}$$
$$OAT = 18 \text{ °F}$$
$$MAT = 69 \text{ °F}$$

Then calculate the percentage of OA as follows:

$$\%OA = \frac{RAT - MAT}{RAT - OAT} = \frac{76 - 69}{76 - 18} = .12 = 12\%$$

The same methodology can be applied to determine leakage through a closed return-air damper, a closed hot- or cold-deck damper, or even face-and-bypass dampers. Although the specific formula would be different, the same general format could be used.

If leakage exceeds 10 percent on conventional dampers or 5 percent on low-leakage types, these dampers should be identified as problem dampers and fixed.

General formulas governing mixed airstreams   In this connection, the following general relationships *always* apply to mixed airstreams regardless of whether they're OA and RA, hot-deck and cold-deck, or face-and-bypass:

$$(T1 \times P1) + (T2 \times P2) = TM$$
$$P1 + P2 = 1$$

where:

T = temperature
P = decimal proportion of total air flow
1 = airstream 1
2 = airstream 2
M = mixed airstream

(*Note:* 1, 2, and M refer to the airstreams to which the temperatures and decimal proportions apply.)

By measuring the temperatures in all three airstreams in a mixing situation, one can then determine the proportions of the airstream desired by simply plugging into these formulas and solving for the missing proportion. (A decimal proportion can be expressed as a percentage by multiplying by 100.)

Action to take

Linkage adjustment   If leakage is significant, your first action is to examine the linkage connecting the damper blades to each other and to the motor. Look to see if some adjustment can be made to close the damper more fully when its motor has driven it to its end position. Such an adjustment is suggested if you see continuous gaps along the edges of damper blades when there's supposed to be edge-to-edge contact.

Gasketing   Some damper blades, though, even when fully closed and pressed together, still leave sizable gaps due to uneven edges or a poor fit between edges. One might consider in this case gasketing the edges to provide a soft, flexible seal that "gives" when the blades are pressed together. Gasketing material is available that can be glued or even bolted or screwed onto the blade edges to provide such a seal.

Verify leakage reduction   Once corrections have been made, readings should be taken to verify that a leakage reduction has been achieved.

SUMMARY OF ACTION STEPS

1. Identify dampers that leak
   a. Inspection (gaps? daylight? cold air?)
   b. Measurement
      Air flow—direct (not recommended)
      Temperatures (best)
   c. Calculate percent leakage
   d. Problem damper if leakage is:
      Over 10 percent (conventional damper)
      Over 5 percent (low-leakage type)

2. Fix problem dampers
   a. Adjust linkage (close gaps)
   b. Gasket edges (flexible seal)

3. Verify leakage reduction (remeasure)

O&M Procedure 3: Minimum Outside-Air Setting Adjustment

What it is   It is common practice to bring outside air (OA) into buildings mechanically, and usually codes or good engineering practice dictate the minimum amounts required. The OA damper controls are then set in such a way that, when the supply fan is on, the OA quantity never drops below that minimum amount (though it may at times go above that amount when it is advantageous for it to do so).

These minimum air quantities are determined by the designer and are incorporated into the construction documents. It is up to the installing contractor—usually the temperature control or the test and balance (T&B) contractor—to make the proper adjustments during original installation so that actual minimum air quantities are set to the design value.

If minimum OA quantities are not listed (in percent or cubic feet per minute) on the construction documents or in a T&B report, it is advisable to contact the original design engineer to attempt to obtain them. (Failing that, it is possible to calculate them based on area or occupancy, but that might be time consuming.)

How to measure minimum OA quantity   This O&M procedure simply requires making a check of the actual minimum OA quantity and, if it differs significantly from the design value, making adjustments to bring it into conformance. The procedure prescribed in O&M Procedure 2 for determining the percentage of OA can be used in this case as well.

Action to take

   Systems with separate minimum OA dampers   In air-handling systems where a separate set of dampers simply opens fully when the supply fan runs, it is necessary to adjust linkage so that the maximum open position of the dampers is limited, thus restricting air flow.

   Systems with minimum OA stop on dampers   In the case where there is one set of OA dampers and minimum OA is determined by damper position—controlled by air pressure or an electric signal passed through a minimum OA position switch—the switch needs to be recalibrated to determine the *exact* position that allows the amount of OA you want. You should check not only switch position but also the exact pneumatic control air pressure to the OA damper motors at minimum. A series of such readings and corresponding

OA amounts will provide you with the information you need to achieve a desired minimum OA quantity or percentage. The minimum air position should be marked right on the switch dial face, along with the corresponding control signal value.

Verification of adjusted OA   In either case, verification of the new setting should be done by taking temperature measurements of the airstreams and then calculating the proportion of outside air, as described in O&M Procedure 2, page 8.

If it is found that an adjustment is warranted, repeat the preceding action.

SUMMARY OF ACTION STEPS

1. Measure *actual* minimum OA (cubic feet per minute)

2. Find *design* minimum OA (cubic feet per minute)

3. Compare *actual* to *design*

4. Identify systems for reduced minimum OA and calculate how much for each

For separate damper section:
5. Adjust linkage to restrict air flow

For minimum stop on single section damper:
5. Calibrate minimum OA position switch
    a. Minimum OA quantity
    b. Control pressure/electric signal

6. Reset switch to reduced air flow (cubic feet per minute)

7. Verify reduction

O&M Procedure 4: Manual Mixed-Air, Cold-Deck, and Discharge-Air Temperature Reset

Why reset is important The purpose of maintaining a low mixed-air temperature (MAT) in cold weather—and similarly a low cold-deck temperature (CDT) in warm weather—is to meet the cooling requirements of *all* spaces served; this means providing a temperature low enough to meet the cooling load of the space with the *most severe* cooling need. Otherwise, that space would over-heat.

In the days before energy availability and cost were problems, it was customary to set these temperatures at a constant level that was sure to be low enough to meet cooling demands at *all* times under the *worst* conditions. The increasing cost of energy in the last 10 years, however, has made it apparent that maintaining excessively low temperatures is a wasteful practice.

To the extent that MAT and CDT—or, in general terms, supply-air temperatures—are lower than they need to be to maintain space conditions, energy is being wasted. In many situations, at conditions that prevail much of the time, these temperatures may be considerably lower than that required by any one space. In order to avoid overcooling, therefore, this air must then be reheated or mixed with warm air to give the required supply-air temperatures. In such situations, *considerable energy can be saved by maintaining higher temperatures*. In these situations, some periodic adjustments can be made manually which, though not perfect, can still result in big energy savings.

What it is What is being referred to here is the manual resetting of CDTs and (on air systems with return air) MATs on a periodic basis (once or even twice a day). The amount that these temperatures can be reset each day (that is, how far you can raise them before you reach the highest temperature that can be maintained while still keeping spaces cool enough) can only be determined by experience and careful evaluation. It depends largely on the kind of weather being experienced and, in some cases, on the kind of cooling loads expected in the spaces served by the systems.

Action to take

Identify systems Systems that are good candidates for reset are ones where, for a good portion of the time, there is a significant difference between *available* and *required* supply-air temperature. While you may be able to identify these systems right away if you're familiar with the areas served, occupancy, and complaint history, you might also make a sampling of air temperatures. Comparing air temperatures entering spaces expected to have the most severe cooling load with air temperatures leaving the cooling coil under several sets of conditions would give you a good indication.

How much to reset and how to schedule The more you know about your systems, the better you're able to determine the reset potentials. If you know, for instance, that a certain system supplies an area that includes a south-facing space with lots of glass and that that space needs the most cooling on sunny days around noon, you might estimate that on overcast days you could still cool that space adequately with a supply-air temperature 5 °F higher than the one you have now. A good program, therefore, might be to reset the MAT 5 °F higher on such days.

In general, deciding on the specifics of a reset program requires consideration of numerous factors—among them space orientation, occupancy patterns, occupant sensitivity, severity and diversity of solar or other loads, and the history of occupant complaints for the spaces served. A gradual, step-by-step approach may be warranted to test the acceptability of a changed mode of operating.

Once determined, the proposed program should be put in written form so that those assigned to carry it out know clearly what it is and so that the effects can be judged in the light of a written record of changes made. Any subsequent revisions should also be put in writing.

Assign responsibility   Reset is best implemented by assigning an individual the responsibility for making the adjustments daily—at the beginning of the second shift, for instance—with the day's new temperature settings to be determined in consultation with the director of physical plant, taking into consideration the variables of weather, load, etc. that would affect such a decision.

Watching the effects   It is then important to keep track of the effects: Are satisfactory conditions still being maintained in the spaces affected? To check this, you would closely observe the conditions in that space, especially around midday, to see if it was still being adequately cooled.

The important thing is to develop a plan, to follow it, to carefully observe the effects, and then to make adjustments to the plan according to those observations.

Savings are significant   Savings can be considerable, depending on the degree of reset possible. A raised average MAT will result in heating energy savings; a raised CDT during the cooling season will result in reduced refrigeration costs. (One must be careful in raising CDT in humid weather: If you overdo it, the cooling coil will fail to remove enough moisture from the air, and spaces will become uncomfortable.) An added benefit of raising MAT in wintertime is reduced humidification requirements in those systems where moisture is being added (resulting from reduced outside-air quantities). See ECM 2, page 109, for some rules of thumb on dollar savings to expect.

SUMMARY OF ACTION STEPS

1. Identify applicable systems
   a. Review areas served, occupancy schedule, complaint history
   b. Measure air temperature entering space, leaving cooling coil
   c. Compare (big difference: good candidate)

2. Decide reset amount (degrees) and reset schedule
   a. Consider:
      Space orientations
      Occupancy patterns
      Occupant similarity
      Severity and diversity of solar (or other) loads
      Complaint history
   b. Consider gradual, step-by-step approach
   c. Commit to writing

3. Assign responsibility to implement

4. Observe effects (rooms hot? discomfort? complaints?)

O&M Procedure 5: Humidity Controls Calibration and Reset

Obviously, this O&M procedure applies only to air systems in which moisture is added to raise relative humidity in spaces served. Low indoor humidities occur primarily either in very dry climates or during cold weather in northern climates. Humidification is practiced mostly in medical facilities, but also sometimes in sophisticated office space and frequently in special-purpose spaces or buildings.

Why it's important   There are two reasons why the calibration of humidity controls is important. One is that, of the dozens and dozens of control devices (including room thermostats) used in a typical air-handling system, humidity controls are some of the few that have a big effect on energy use. The other is that humidity control devices are very likely to go out of calibration quickly.

Taking humidity readings   The definition of relative humidity and the relationship between air temperature and relative humidity are discussed in the section entitled "Humidification Equipment" which begins on page 38. If humidity control devices are to be calibrated, it is necessary for the operator to measure and/or derive actual relative humidity from a chart or table.

Sling psychrometer   There are numerous devices on the market that read humidity; probably the most dependable of the inexpensive, hand-held types is the sling psychrometer. See the section entitled "Humidity Measurement" which begins on page 127 for more details on this device.

Where to measure humidity   The best place to measure the humidity of an airstream is in the duct—the same place the humidity controller senses it. It would be best if you could enter the return duct through an access door and hold the psychrometer as the air moved by it. Most return ducts, however, do not allow this kind of access.

Sticking the psychrometer into the duct through an access door or another opening can result in false readings if there is leakage around the opening through which the device is inserted. (Such leakage can allow mechanical room air to enter and possibly contact the device's bulbs.)

In that type of situation, it would be better to take several readings of conditions in typical *spaces served by the air-handling system* and average those readings. You would then be assuming that the air in the return duct was at the same humidity as in the spaces. This is a reasonable assumption (unless the return duct leaks a lot), and you can still make a useful calibration adjustment since it is the space humidity that you ultimately want to control anyway.

How to proceed

(a) The instrument is whirled around for the prescribed minimum time, usually 1 to 2 min. (Follow the manufacturer's recommendations.)

(b) Readings of both thermometers are taken, preferably as soon as possible. The uncovered dry thermometer reads dry-bulb temperature while the wet, wick-covered thermometer reads what is called wet-bulb temperature. The wet wick on the bulb, when slung through the air, has a cooling effect and depresses the bulb temperature. The amount of temperature depression below the dry-bulb reading depends on the amount of moisture in the air. If the air were 100 percent saturated (that is, if it were like fog), the depression would be zero, and 100 percent relative humidity would be indicated. If the air were dry, the depression would be considerable, indicating a low relative humidity.

(c) With the wet- and dry-bulb temperatures recorded, you then consult either the psychrometric chart or a table derived from that chart. (A typical psychrometric chart is illustrated in Figure 1-2.) By locating the lines corresponding to the two temperature readings (dry-bulb on the bottom, wet-bulb along the curved line on the left), you find the point where they intersect. You then can read relative humidity (RH) by estimating the location of that point with respect to the constant RH lines, also shown on the chart.

(d) If the air is not at the RH you want (say, if RH is too high), adjust the humidity controls. Then repeat the cycle of measure and adjust until you read the desired RH—or at least get within 5 percent of it. Record the readings as you go along. Then mark the dial of the control with the corresponding humidity level. Identify what humidity levels (RH) you wish to maintain, and compare them to the actual levels.

(*e*) Since humidity controls do go out of calibration easily, the procedures should be repeated frequently—like every month during the humidification season.

SUMMARY OF ACTION STEPS

1. Identify systems with humidification (moisture being added)
2. Obtain appropriate humidity-measuring instrument (sling pyschrometer ok)
3. Decide where to measure (in duct or in space)
4. Identify desired levels (percent RH)
5. Read and record actual RH
6. Compare; if too high, reduce humidification rate (via control setting)
7. Repeat (4) and (5) as needed until desired level obtained
8. Mark dial of humidity control with position and value (percent RH)
9. Recheck calibration frequently

Other Measures

The foregoing O&M procedures are considered to be ones that can save the most energy, but they are not necessarily the only ones.

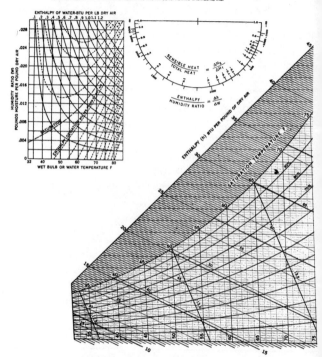

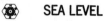

FIG. 1-2 Psychrometric chart; relative humidity can be found by locating point where wet- and dry-bulb temperature lines cross. (Chart courtesy ASHRAE.)

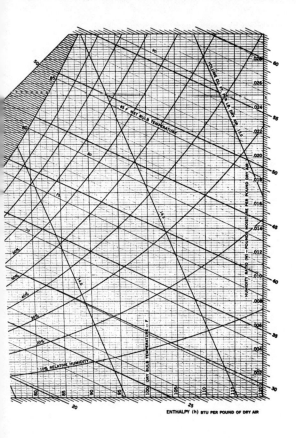

ENTHALPY (h) BTU PER POUND OF DRY AIR

There are some others that should be mentioned as well. While not producing energy use reductions as large, they nevertheless are basic to good system operation and maintenance and will, over the long run, result in more efficient system operation.

O&M Procedure 6: Patch Ductwork Leaks

One of these other O&M procedures is inspecting the ductwork and distribution system to see if there is air leakage and then taking steps to correct such leakage if found. Small air leaks can often be patched with caulking compound or properly placed duct tape. Larger gaps or holes, however, may require sheet metal patches (sometimes gasketed) held on with sheet metal screws.

Substantial leakage from supply ducts means that some air being pumped by the fan to supply occupied spaces is not reaching those spaces. Therefore, either the spaces receive insufficient air or the fan is pumping more air than necessary, and thus consuming extra energy.

O&M Procedure 7: Periodic Filter and Coil Cleaning

Another procedure—this one is more a maintenance than an energy-saving item—entails ensuring that filters are changed periodically and coils kept clean. Even though virtually all coils on air-handling systems are theoretically protected from dirt by filters upstream, filters are not perfect and some particles do get by. In time, coils, with their closely spaced fins, will collect dirt on the upstream side and tend to clog. This not only increases air flow resistance, it reduces heat transfer capability as well. Most maintenance programs include periodic inspection and changing or cleaning of filters. A similar procedure should be set up for inspecting and vacuum cleaning the upstream side of all coils on a somewhat less frequent basis (every 3 to 6 months).

O&M Procedure 8: Periodic Controls Calibration

Finally, though it should not have to be mentioned, controls are the "brains and nervous system" of an air-handling system. If they get out of calibration or malfunction, chances are that some degree of inefficiency in system performance will result. Most existing controls for air-handling systems are pneumatic, and air used in pneumatic systems does contain dirt and moisture (despite filters and dryers). Therefore, pneumatic control devices will in time get out of calibration—perhaps faster than one might expect.[1]

Although many building managers have maintenance contracts with outside firms who periodically calibrate controls, others have in-house staff capable of doing simple calibration tasks. Controls in general should be checked and calibrated at a minimum interval of 6 months. Airstream temperature controllers (such as mixed-air, cold-deck, hot-deck) should be checked much more frequently (once a month wouldn't be too much) since their accuracy has such a big effect on energy efficiency.

[1]Douglas C. Hittle, William H. Dolan, Donald J. Leverenz, and Richard Rundus, "Theory Meets Practice in a Full-Scale Heating, Ventilating and Air-Conditioning Laboratory," *ASHRAE Journal*, November 1982.

2

Description of Air-Handling System Components

A building's air-handling system might be thought of as a circuit, or a closed loop: Air is distributed from a central point to occupied spaces and then brought back for reuse or exhaust to the outside; it travels in a loop. During its circuit around the loop, various things happen: It may be cleaned, heated, cooled, humidified, dehumidified, pressurized, speeded up, slowed down, and diffused. The collection of components that causes air to circulate in this manner and to undergo these processes is called an *air-handling system*.

The purpose of this chapter is to identify and describe the major components of air-handling systems, including their characteristics, functions, and other key facts about them.

The major components of air-handling systems are: intake, exhaust, and relief openings; automatic dampers; filters; humidification equipment; coils; the bypass section; fans; the distribution system and ductwork; housings, enclosures, bases; vibration isolation and sound control appurtenances; motors and drives; and automatic temperature controls. Each of these is described in a separate section. The section order is the same as the component order one might find in a typical system, moving downstream from the intake. Figure 2-1 and subsequent illustrations depict the components.

INTAKE, EXHAUST, AND RELIEF OPENINGS

Purpose

The purpose of intake, exhaust, and relief openings is to let air into or out of a building. Although small structures like homes do not require one, it is necessary in larger buildings to design into the air-handling system a positive means of drawing in outside air. It is good engineering practice to do so, and besides, building codes—and sometimes special purpose building requirements—demand it.

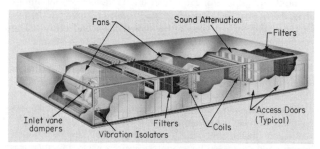

FIG. 2-1 Cutaway of package rooftop air-handling unit shows some typical components. (Diagram courtesy Buffalo Forge.)

What They're Like

Most openings for the intake, exhaust, or relief of air in buildings have vertical fixed-blade louvers with the blades arranged to minimize entry of the elements. See Figure 2-2. (Some blades have a special design to help them do this; these blades are referred to as "weatherproof.") Another configuration for building openings is the rooftop device, with either a gooseneck or a mushroom type housing.

FIG. 2-2 Outside air intake louver. (Photo courtesy Carnes.)

Relief openings are used in a building where there is no positive means of exhaust; air is then allowed to force its way out. In older buildings there may be enough cracks, leaks, and other holes so that excess air brought in will be able to get out, but in tighter buildings openings are needed to let air out. For example, take a system where a single fan is used to both supply and return air: When the system is operating with a high percentage of outside air, there is no place for that air to go in the building except out. Relief openings provide a path.

Relief openings are also used to let outside air come in to replace exhausted air. A common example of such a use would be in a small boiler room where combustion air enters the room through a louvered relief opening.

Birdscreen

It is customary to cover air openings with a birdscreen. This should not be any less than $\frac{1}{4}$ in. mesh screen, usually located on the inside face of the louver. By no means should insect screen be used because it will very quickly clog up with dirt and block the opening.

Pressure Balance

What it is The amount of air introduced and exhausted from a building affects the building's pressure balance. Very simply, a building is under positive pressure if more air is brought in than exhausted: Air would tend to flow *out* an open window. If the opposite is true and air tends to flow *in*, the building is under negative pressure.

In examining the question of pressure balance, one must consider *all* the intakes and *all* the exhausts. For example, for an air-handling system serving a certain area, one must consider not only the air supplied and exhausted by the main system but by all *other* exhaust systems serving that area as well, such as toilet room or storage exhausts.

Building versus room pressure balance One must distinguish the pressure balance of an entire building from room pressure balance. The latter is governed by the amounts of air brought into and exhausted from *a given room;* the balance in the room may or may not be the same as the overall building pressure balance. In other words, an entire building may have a positive pressure, but a given room a negative pressure. Room pressure balance is very important in certain specialized buildings like labs and hospitals for controlling the spread of toxic or objectionable fumes, contaminated air, or bacteria. See Table A-1 in the Appendix, which shows some recommended pressure balance requirements for typical hospital spaces.

AUTOMATIC DAMPERS

What They Are and What They Do

Automatic dampers are generally a series of pivoting blades placed in a duct or across an opening; when in a position perpendicular to air flow, they close off the passage and therefore the air flow through that duct or opening. Although dampers can be manually moved (for air-balancing and adjustment purposes, for instance) to permanent settings, when used for control they are usually automatically operated by damper motors. The blades of a damper assembly are usually linked together by levers and linkages (but sometimes by gears), and a shaft from a damper motor is connected to one of the blades to drive the entire blade assembly to the desired position.

Thus, the purpose of automatic dampers is not only to cut off air flow but also to throttle (vary) it.

Types of Damper Motors

Damper motors are of two types: pneumatic (driven by air pressure) and electric. Pneumatic damper motors are usually equipped with a spring to keep the damper in one extreme position or the other. This is called a motor's "normal" position. (The same term applies to valves.) This is always the position to which a damper or valve will travel when there is no air pressure or no power. In the case of outside air dampers, the normal position is always the closed position.

Damper Flow Characteristics

Dampers are further classified as either opposed-blade or parallel-blade. See Figure 2-3. *Opposed-blade* means that the blades pivot in opposite rotational directions; *parallel-blade* means they pivot in the same direction. Although most automatic dampers in common use are the opposed-blade type, parallel-blade dampers do have better air flow characteristics.

Flow characteristics describe how much air passes through a damper as it travels from fully closed to fully opened. Ideally, for instance, a damper in the 30 percent closed position (by angle) should pass only 30 percent of total air flow. Unfortunately, this is not the case, but parallel-blade dampers do more closely approach this ideal.

Where Automatic Dampers Are Used and Why

Outside, return, and exhaust air   Dampers can be used in several locations and for various purposes in air-handling systems. Already

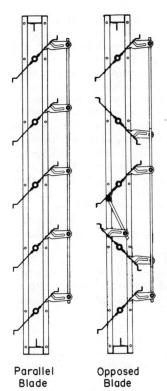

Parallel
Blade

Opposed
Blade

FIG. 2-3 Cross section of parallel-blade and opposed-blade movable dampers; note linkage differences. (Diagram courtesy American Warming and Ventilating.)

mentioned is their use in outside- and exhaust-air openings. Another related location is the recirculated-air duct which carries air returned from the spaces to mix with outside air. Outside- and return-air dampers generally work in tandem (that is, when one is open, the other is closed) with the exhaust-air damper closely paralleling the operation of the outside-air damper.

It is common, particularly with larger fan systems, to have two sets of dampers for control of the outside-air intake, one set for "minimum" air, and the other for the rest. Although this requires another set of damper motors and division of the opening into a minimum OA section and a "maximum," it allows closer balancing of OA quantities and generally better control of air brought in.

The alternative way to control outside air is with a single damper with a minimum stop (or position setting). Once opened, the damper never closes below that setting. Since a single damper is mostly closed at this minimum point of operation, its flow characteristics are not favorable, and it is difficult to obtain the correct OA quantities.

Bypass air   If the fan system is equipped with a bypass section, that section will need to have face-and-bypass dampers which direct air either through the coil or through a bypass around the coil. See Figure 2-4.

Hot deck and cold deck   A common use of dampers is on a fan's discharge as hot-deck and cold-deck dampers on multizone units. (Such a unit is described in the section entitled "Air Mixing: Multizone," which begins on page 97.) A set of these dampers is required for each zone of a multizone unit, with both the hot- and cold-deck blades mounted on the same shaft. The two are designed to work in tandem, and their purpose is to blend cold air and hot air to meet the needs of a particular zone being served.

FACE - AND - BYPASS CONTROL

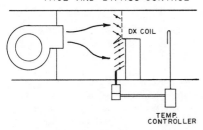

FIG. 2-4 Face-and-bypass dampers allow air to pass through coil, go around it, or some combination of the two.

Shutoff or sectionalizing Sometimes an air-handling system's distribution network will be equipped with automatic shutoff or sectionalizing dampers, used to keep air from going to one portion of the building. Such dampers are generally of the two-position type (open or shut).

Inlet vane or vortex Fans themselves may be equipped with dampers to adjust the amount of air flowing outward from them. This output is always measured in cubic feet per minute (ft^3/min) of air flow. On centrifugal fans, these are *inlet vane dampers* (sometimes called *vortex dampers*), which are radial blades that pivot. They are the most common means of making final air balance adjustments to total fan output and are frequently used as the automatic control device for variable volume air-handling systems. Although it is possible to adjust the quantity of air handled by a fan with discharge dampers, it is more efficient to do it with inlet dampers.

Why Leaky Dampers Waste Energy

An increasingly important characteristic of dampers—OA dampers in particular—is their closure condition: that is, how much they leak when they are supposed to be fully closed. This is important because *supposedly closed dampers that leak either: (1) allow unheated (or uncooled) air to enter the buildings when fan systems are stopped; or (2) allow warm air to mix with cold air, causing unnecessary energy use.* How tightly the edges of damper blades fit when closed is therefore very important. Damper blades are often fitted with gaskets or seals along their edges in order to block off openings that could leak. See O&M Procedure 2, page 9.

Closure condition of mixing dampers is also important—particularly those dampers that mix heated and cooled air, because obviously leakage of heated air into a cold airstream warms the cold air being supplied. You're therefore wasting any energy used to heat or cool either stream. Modern damper designs have given greater consideration to low leakage design.

FILTERS

Purpose

The purpose of filters is to prevent dirt from entering a circulating airstream and to remove dirt from it. If air-handling systems did not have filters, dirt, in addition to being blown into the occupied spaces, would accumulate on coils, ducts, and other system elements, and eventually some system elements—like coils—would clog up and severely restrict air flow. This in turn would eventually reduce the air-handling system's heating and cooling capacity.

How Filters Are Classified

Filters are classified in several ways.

Throwaway or cleanable   Today throwaway filters are common in both small and large air-handling systems and are usually more cost effective to use than cleanable filters for low-efficiency filtration.

Efficiency   Filters are also classified according to their efficiency at removing particulate matter, with those capable of removing fine particulate matter classified as high-efficiency. See Figure 2-5. Filter efficiencies are designated as a percentage, with the high-efficiency types ranging from 85 percent to close to 100 percent efficient.

Arrangement   There are several common ways that filters are placed in an airstream.

Perpendicular   Thin, low-efficiency filter sections are flat and come mounted in frames about 4 ft² or so in area. These sections are inserted in a holding frame (filter bank), which often is perpendicular to or directly across the air flow. High-efficiency filters, because of their greater depth (12 in.), use this arrangement too, though such filters are located downstream of the fan (on the pressurized side, to avoid drawing in dirt).

Angled   The filter bank may be of sawtooth arrangement, so that each filter section is at a 45° angle to air flow. See Figure 2-1. This is done so that the entire filter bank takes up less cross-sectional area of the housing. Thin, low-efficiency filters are frequently arranged this way too.

Bag   A common filter bank configuration is a series of bags which

FIG. 2-5   High-efficiency particulate air (HEPA) filter; efficiency of removal of particulates 0.3 microns and larger generally ranges from 85 percent to close to 100 percent. (Photo courtesy Koch Filter.)

blow out downstream and provide a lot of filter medium area per cross-sectional area of airstream. See Figure 2-6. Figure 2-1 also depicts bag filters in place. The bags, each about 2 ft × 2 ft, are held in place by a frame mounted across the duct cross section.

Roll   A roll filter consists of a big spool of thin, low-efficiency filter medium that traverses slowly across the filter housing, thereby moving the dirty portion to a second spool and introducing clean filter medium at the other end.

FIG. 2-6 High-efficiency (bag) filter; bags blow out downstream, provide a lot of medium area for the airstream cross section. (Photo courtesy American Air Filter.)

Filtration method   Finally, there are filters that rely on chemical action as opposed to mechanical action for their cleaning process. One such type in common use is the activated charcoal type; certain air impurities are absorbed in the charcoal, which then must be periodically removed and renewed. Though chemical filters are more expensive to buy and maintain, some codes relax their high outside-air requirements when such filters are used.

Maintenance Considerations

By their very nature, filters require frequent attention. How often air filters must be replaced or cleaned depends on how fast they get dirty. This in turn depends of course on how dirty the air is.

The frequency of periodic maintenance is based on experience, but there are devices that give an indication or automatic signal when filters are dirty.

A common device on large filter banks is a differential static pressure gauge, which indicates a filter is dirty when the pressure drop across it reaches a high value.

HUMIDIFICATION EQUIPMENT

While control of humidity levels may be done strictly for reasons of comfort in conventional buildings, in certain others (like hospitals) it is considered essential. Although humidity control involves the removal of moisture during periods of excess as well as the addition of moisture during periods of deficiency, the main discussion here will concern the addition of moisture. Removal of excess moisture is part and parcel of the cooling process, and you can usually achieve good humidity control in humid climates or during summer months (when excess humidity can be a problem) if you maintain good control of the cooling process.

Why Humidity Is Important in Hospitals

There are two reasons why humidity control is so important in hospitals: (1) Respiratory distress in patients is minimized at levels approaching 50 percent relative humidity[1]; and (2) static electricity sparks, which present a danger of explosion where pure oxygen is used, are eliminated at higher humidity levels. Further, there

---

[1]"ASHRAE Standard 55-1981, Thermal Environmental Conditions for Human Occupancy" (ANSI B193.1-1976), American Society of Heating, Refrigerating and Air-Conditioning Engineers, Inc., Atlanta, Georgia, 1981.

seems to be a relationship between the proliferation of certain bacteria and humidity levels, but studies are inconclusive as to exactly what that relationship is.

What Is Relative Humidity?

Air inside a building tends to be dry in cold weather because the relative humidity of air tends to drop as it is heated. A key fact about moisture in air is that the warmer the air is, the more moisture it can hold. *Relative humidity* is the ratio of how much moisture air holds to how much it *can* hold at a given temperature. Relative humidity is expressed as a percentage.

Why Moisture Is Added

Outdoor air in wintertime may contain a normal amount of moisture and thus may have a relative humidity in the range of 40 to 60 percent. However, if that air is brought up to room temperature without any mosisture added (which is what happens when a fan system heats it), its relative humidity will drop considerably—more the colder it is. (See previous paragraph.) Therefore, if you want to maintain relative humidity levels of around 50 percent, you *must* add moisture to the indoor air.

The humidity requirements in any building during cold weather are further increased as larger amounts of outside air are introduced—for the same reasons cited above.

How Humidification Is Done

Although it is possible to humidify on a space-by-space basis using domestic type humidifiers or vaporizers, on a larger scale humidification is best done centrally through the air-handling system.

The object is to add moisture to the airstream to a sufficient extent to raise the general humidity level in the occupied spaces to around 40 to 50 percent.

There are several means of adding moisture to an airstream. These are: to discharge steam, to evaporate water, and finally to spray in fine water droplets. (The latter process is called *atomization*.) Each of these is described briefly here.

Steam grid   The most common technique for adding steam to an airstream is through a steam grid, which is basically a tube with holes in it through which steam can escape. See Figure 2-7. The grid—or a series of grids—is placed across the cross section of the airstream, and steam is ejected through the holes. The steam supply is controlled by an automatic valve which responds to measured humidity levels in either the occupied space or the return airstream.

Evaporative methods

Heated water pan   A second method of airstream humidification is to evaporate water into the airstream via a pan equipped with a heating coil and mounted on the bottom of a duct. The heat speeds up the evaporation process.

Sprayed coil   A method frequently used for very close humidity control is the sprayed coil; in this method, recirculated water is sprayed on a coil and a dewpoint controller maintains the proper humidity level through the cooling coil control valve. The sprayed coil was popular on older systems.

Air washer   An air washer adds humidity by running the airstream through a deluge of sprayed water followed by a means of eliminating carried-over water droplets.

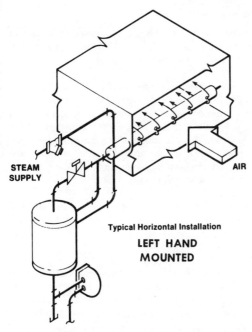

STEAM
SUPPLY

AIR

Typical Horizontal Installation

LEFT HAND
MOUNTED

FIG. 2-7 Steam grid humidifier; horizontal element (grid)
extends across airstream and steam discharges from holes.
(Diagram courtesy Spirax Sarco.)

Atomizer A third method is to use a device that atomizes water
and ejects it into the airstream as very tiny droplets, which then
evaporate.

Humidification Problem

A common problem with humidification is caused by impurities (minerals) in the water used. The nature of the problem is described below for each humidification method.

Steam   In the case of steam, this problem occurs back at the steam generator, whether that be a local device or a central boiler plant. Steam used for humidification does not return to the boiler as condensate—and the steam generator therefore requires water makeup. To the extent that the water makeup contains impurities, chemical treatment processes are needed to avoid corrosion and buildup of scale and deposits on boiler surfaces. More makeup means more impurities, and thus more treatment.

Evaporated water   In the case of water pan humidification, the water evaporated from the pan leaves its residue in the pan itself or on heating pipes in the pan; again, scale, deposits, and corrosion can eventually cause problems.

Atomization   The problem with impurities is not avoided with the atomization method either since, in this case, the impurities are sprayed into the airstream along with the water itself; these impurities eventually settle out as dust, either in the ductwork or in the spaces served.

Sprayed water   Sprayed coil assemblies, of course, are left with water residue and corrosion on the coil being sprayed and on the nozzles dispensing the spray. The best means of minimizing the residue and corrosion problem due to water impurities in systems such as the sprayed coil is simply to keep the system well flooded to avoid letting the wet elements dry out. This tends to wash the impurities down the drain before they can dry and cake up on a surface.

Air washers, by their very nature, are always deluged when they are in opeation.

COILS

What Are Coils?

Coils are the main devices for transferring energy to or from air-streams in an air-handling system. Energy is transferred by the addition of heat (sensible energy) and by the removal of heat and moisture (sensible and latent energy). These processes, of course, are called *heating* and *cooling,* respectively. (Moisture removal itself is called *dehumidification.*)

What Coils Consist Of

Coils used in most heating and air-conditioning applications consist of a series of tubes assembled in a casing. Thin plates, called *fins,* are mechanically bonded (that is, pressed on) to the tubes. The purpose of the fins, of course, is to expose the maximum amount of conductive surface to the air flow without causing excessive blockage of air and hence resistance to the flow.

Importance of Fin Spacing

The spacing of fins is important because it affects both energy transfer and flow resistance. (This spacing is usually expressed in terms of number of fins per inch along a tube.) Obviously, the closer the fins are spaced together, the more surface exposure to the air there is and the less bypass air (air that passes through without contacting the fins) there is. However, the closer the fins, the more air friction there is, and the higher the pressure drop across the coils. Fin spacing generally runs 8 to 14 fins per inch in most applications. A good rule of thumb for general HVAC applications to avoid too-frequent clogging is 8 fins per inch.

Heat Transfer Media

The medium used in the coil is either a hydronic fluid, steam, or a volatile refrigerant (like Freon). Hydronic fluid can be either hot water, chilled water, or some type of antifreeze mixture (like ethylene glycol). Refrigerant, of course, is used with direct expansion (DX) air-conditioning systems where no chilled water is involved. Some codes, however, restrict the use of DX systems in hospitals (Chicago code, for example).

Where Different Types of Coils Are Used

Hydronic (Figures 2-8A and 2-8B)   Most large modern HVAC comfort air-conditioning applications utilize hydronic coils for energy transfer. The use of ethylene glycol in chilled-water coils has the advantage of making it most unlikely that the coolant will freeze solid and thus burst the coils open; however, ethylene glycol's heat transfer characteristics are inferior to those of plain water, and, of course, filling an entire system or replacing fluid lost by leaks with ethylene glycol is expensive. To avoid freeze-up, many operators drain their chilled-water coils during the cold months and temporarily fill them with ethylene glycol to ensure that all water is driven out.

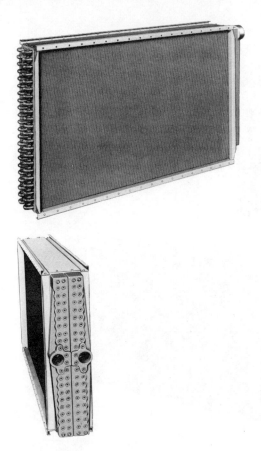

FIG. 2-8A Double-row, drainable serpentine water coil: two views. (Photos courtesy Trane.)

FIG 2-8B Four-row water cooling coil with same end connections: two views. (Photos courtesy Trane.)

Steam (Figure 2-9) Steam coils are used extensively on older systems where steam is available anyway—and particularly on preheat coil applications, where the coil is subject to extreme cold temperatures. Steam is not generally as suitable for applications where precise leaving air temperatures are desired because steam is difficult to control and because it has certain other inherent drawbacks.

Direct expansion (DX) (Figure 2-10) Most small air-conditioning units are equipped with DX coils, which avoids the necessity for installing or tying into an existing chilled-water system. Also, freeze-up problems inherent in a water system are avoided.

Coil Controls

Control of the amount of energy transfer through coils is generally accomplished by varying the transfer medium flow rate, the transfer medium temperature, or the amount of air passing over the surface.

Hydronic On hydronic coils, throttling valves or bypass valves are used to vary flow through the coil, either by restricting flow or by diverting it around the coil. Hydronic medium temperature is usually varied by means of a mixing valve which blends new entering water with return water.

Steam Steam coils are generally controlled by a throttling valve, although it is difficult to maintain good cross-sectional steam distribution and even heating with this kind of control. In the case of preheat coils, the valves are generally controlled so that, below a certain outdoor temperature (like 30 °F), they remain open constantly at full flow.

FIG 2-9 Single-row heating coil with opposite end connections: two views. (Photos courtesy Trane.)

FIG. 2-10 Direct expansion coil; note the distribution manifold for the incoming refrigerant. (Photo courtesy Trane.)

Direct expansion (DX) DX coils are controlled by a thermostatic expansion valve which regulates the amount of refrigerant flow according to pressure in the suction line (which contains the cold refrigerant entering the coil).

Heating Coils

A heating coil's function is to add sensible heat to air, bringing it from an entering temperature to a desired leaving temperature. Heating coils are used in comfort heating applications for preheating, for general purpose heating, or for reheat (or "booster") purposes.

Preheat coils   Preheat coils are used on 100 percent outside-air systems or on air systems where the expected mixed-air temperature is relatively low (below 50 °F). Preheat coils should be sized to heat incoming air to 35 °F or more, and steps should be taken to prevent freeze-up since the opportunity is certainly there. With steam coils, the design must be of the nonfreeze type where the condensate is self-draining. In the case of hot water coils, antifreeze fluid is frequently used, or else continuous water flow is maintained through the coil whenever temperatures are below freezing. Preheat coils are sometimes shut off (via the controls) above 45 °F outside-air temperature to avoid overheating in the rest of the system.

General purpose heating coils   Coils used for general heating purposes are located fairly close to the air-handling unit—upstream of it on draw-through systems. An example of a typical application is a single-zone heat-cool-off type system (see Figure 3-1 on page 91) where the air is either heated or cooled in sequence, depending on the needs of the spaces being served. Downstream of the fan, general purpose heating coils usually appear in the hot deck of multizone or double-duct systems, where the coils are the only source of heat for the system. See Figures 3-3 and 3-4 on pages 95 and 96. (In systems where the lowest mixed-air temperature is expected to be relatively high, it is not necessary to provide a preheat coil.)

Reheat coils   Reheat or booster coils are used in terminal reheat air-handling systems. See Figure 3-2 on page 93. Such coils tend to be small, hydronic, and located in the ductwork close to the occupied space being served. They are always controlled directly from the room thermostat, providing the final "trim" control needed to meet space needs.

Cooling Coils (Figures 2-8A and 2-8B)

Function Cooling coils have an additional function in that they remove not only sensible energy, but latent energy (that is, energy due to moisture) as well. This is a very important role that cooling coils play, especially in humid seasons or climates.

Media Media used in cooling coils are water, ethylene glycol, and refrigerant.

Bypass factor A particularly important characteristic of cooling coils is the amount of air that "bypasses" the coil—that is, the amount of air that, in passing *between* the fins, does not contact them and is thus not reduced in temperature or humidity. A coil is rated with a bypass factor under certain performance conditions, expressed in terms of a percentage. This factor can range from 30 percent for coils with wide fin spacing, high velocities, and few rows to 2 percent or less for coils with close fin spacing, lower velocities, and more rows.

BYPASS SECTION

What It Is

Air-handling systems sometimes include a bypass section as a means of controlling air temperature. See Figure 2-4. This *bypass section* is a prefabricated section of sheet metal that allows air to bypass an associated heating coil or cooling coil, either in whole or in part, and thus not be subject to energy transfer. The bypassed air then mixes with the air going through the coil on the downstream side, and the resulting mixture moves on through to the rest of the system.

Bypass Dampers

Bypass sections are equipped with face-and-bypass dampers which control the amount of air bypassed to meet the required leaving air or space conditions. The dampers operate in tandem; as one opens, the other closes.

Baffle Plate

Because air that bypasses a coil generally prefers to travel the path of least resistance, it is often necessary to place some restriction (usually a perforated plate, called a *baffle*) in the bypass air path to equalize the resistance and thus prevent the volume of air handled by the system from increasing when on full bypass.

Humidity Control Limitations

When the coil in a face-and-bypass section is a cooling coil used for moisture removal as well, bypassing air may result in high-humidity problems in the occupied spaces even though the dry-bulb temperature requirements are met.

FANS

What a Fan Does

A key part of any air-handling system is the fan, since the fan is the device which causes air to move. It does this by pressurizing the air, that is, causing a pressure difference, and thus causing the air to move (through the ductwork) from the area of higher pressure (outlet) to the area of lower pressure (inlet). The part of the fan that rotates and actually causes air to move is called the *rotor* or *impeller*.

Air-Handling Units versus Fans

Sometimes the term *air-handling unit* is used. This is a more encompassing term that refers to the prefabricated portion of an air-handling system, the primary purpose of which is air movement, but which also may consist of several components in addition to a fan. A typical example is a prefabricated multizone unit. A multizone unit, as purchased and delivered to a site all assembled as an integral unit, includes not only a fan but also a cooling coil, heating coil, dampers, and a housing or sheet metal enclosure.

Figure 2-1 illustrates an air-handling unit; Figures 2-1 through 2-17 illustrate fans.

Fan Classifications

Fans are classified in several ways. See Figure 2-11 for diagrammatic illustrations of most of these classifications.

How air is moved (centrifugal, axial)   The most common classification is according to how air is impelled (that is, how the device actually makes the air move). Fans are referred to as centrifugal and axial. A *centrifugal fan* (see Figure 2-12) causes air to move by drawing it in axially (from the side); then the air is moved outward (radially) by centrifugal force as the impeller rotates. An *axial fan* (see Figure 2-13), on the other hand, impels the air along the direction of the impeller axis in the manner of a propeller. See also Figure 2-14.

Type	Impeller Design	Housing Design	Performance Curve
Airfoil			
Backward Inclined Backward Curved			
Radial			
Forward Curved			

CENTRIFUGAL FANS

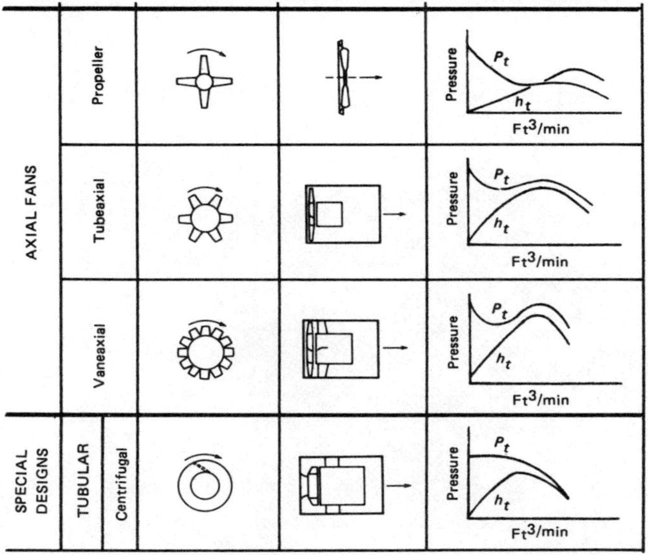

h_t = Total efficiency
P_t = Total pressure

FIG 2-11 Various types of commonly used fans are illustrated; different impeller and housing designs result in different levels of performance. (Diagram courtesy ASHRAE.)

FIG. 2-12 A centrifugal fan with a scroll type housing and backwardly inclined impeller blades; air enters from left, motor and drive are in housing on right. (Photo courtesy ILG.)

Housing design (scroll, tubular) Centrifugal fans are enclosed in a scroll type housing whose shape corresponds to the motion of the air off the impeller and its exit in a direction perpendicular to the impeller axis. The housing configurations of axial fans are often tubular, corresponding to the flow of air along the axis of the

FIG 2-13 A belt-driven axial flow propeller fan; this type of fan is usually mounted in a wall opening. (Photo courtesy ILG.)

impeller. Figure 2-11 illustrates different housing designs diagrammatically, while Figures 2-12 and 2-15 contrast the two housing designs. Figure 2-16, on the other hand, illustrates a tubular centrifugal fan—axial flow, but a centrifugal fan proper.

Pressure range (high-pressure, low-pressure) Air-handling systems are referred to as high-pressure, low-pressure, or medium-pressure, and the fan must be capable of moving the air at the

FIG. 2-14   Another type of belt-driven propeller fan; this one is commonly roof-mounted. Housing shown in raised position. (Photo courtesy ILG.)

pressure the ductwork imposes. High pressure is pressure in excess of 6 in. water gauge (WG), while low pressure is 2 in. WG or less. Medium pressure would be in the range of 2 to 6 in. WG. High-, medium-, and low-pressure ratings for fans are equivalent to Class III, II, and I fan ratings respectively.

Blade configuration (forward-curved, backward-curved, radial)   Blade shape is an important determinant of a fan's performance and efficiency, and so this characteristic is frequently used

FIG. 2-15 A tubeaxial fan: Fan is propeller type, and air flow is along fan axis. (Photo courtesy ILG.)

in describing a fan. Figure 2-17 illustrates radial blades, Figure 2-12 backward-inclined. Figure 2-11 diagrams the differences between the three types.

A variation of the backward-curved is the *airfoil*, which, rather than consisting of a single sheet of uniform thickness metal, is made with two thicknesses of metal to form an airfoil shape.

Fan Drives and Motors

Most fans are driven by electric motors, and, except for small exhaust fans and propeller fans, a belt drive is used to connect fan to motor. The motor can be mounted either on the fan housing or on the base, with provisions for moving the motor to adjust belt tension. Using belt drives makes it fairly easy to adjust fan speed since the belt sheaves can be changed or, in the case of a smaller

FIG. 2-16 A tubular centrifugal fan, including exploded view; note that, although there are centrifugal fan blades inside, overall flow is axial. (Photos courtesy ILG.)

FIG. 2-17 A blower fan. This type of fan is used in low-quantity, high-pressure applications; note radial blades, direct motor drive. (Photo courtesy ILG.)

sheave, adjusted in diameter. Figures 2-13 and 2-17 show belt-drive and direct-drive fans respectively.

Fan Performance Characteristics

How they are expressed Every fan has particular characteristics of performance, and it is important to understand generally what the nature of these characteristics is. *For every fan size within a given type, there is a series of fan curves which describes the performance of that fan.* There are curves for pressure, efficiency, and horsepower, all plotted against fan output (expressed in cubic feet per minute). Fans are factory tested to establish the curves, and a fan installed in the field can be expected to operate on the preestablished curve between shutoff conditions (no flow, maximum external resistance) and free delivery (maximum flow, no external resistance).

Why they are important By making certain key measurements on an air-handling system and/or a fan proper, a user can determine at exactly what point on its curve a fan is operating. This is helpful in determining the effect of adjusting a fan's operation. As mentioned, different fan types exhibit different characteristic fan curves. Some fan curves common to comfort air-conditioning applications are illustrated in Figure 2-11.

Pressure–air quantity curve The most common curve describing fan performance is the pressure–air quantity (ft³/min) curve. See Figure 2-18. This curve is derived by the manufacturer in accordance with the strict requirements of the American Society of Heating, Refrigerating, and Air Conditioning Engineers (ASHRAE) and the Air Movement and Control Association (AMCA). It is derived by plotting total pressure against volume flow rate, that is, air flow expressed in cubic feet per minute. Pressure in a fan system is indicative of energy imparted to the airstream by the fan. This energy can take two forms: static and kinetic.

Static, velocity, and total pressure *Static energy* is that energy needed to actually compress the air and raise its pressure. *Kinetic energy*, on the other hand, is the energy possessed by the air due to its motion. Kinetic energy is indicated by velocity pressure. *Static pressure and velocity pressure together make up a fan's total pressure, which is the true indication of all the energy imparted to the*

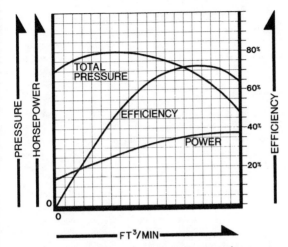

FIG. 2-18 A typical fan pressure–air quantity curve; the curve predicts how a fan will perform over its full range of operation.

airstream by the fan. Static pressure can be measured, as can total pressure, and static pressure is subtracted from total pressure to get velocity pressure.

Static and velocity pressure are interchangeable; that is, if an airstream slows down, the energy (pressure) it had due to its velocity is converted partly to static energy (pressure). The reverse is true if an airstream speeds up. It is important to remember this point, especially when looking at high-pressure systems.

If you measure static pressure at two points in an airstream, let's say at the inlet and the outlet of a fan, the pressure difference between those two points is the same as the total pressure difference—and thus indicative of the energy added to the air—*only* if the "in" and "out" velocities of the air are identical. The same is true for pressures measured at any point in an air-handling system.

Fan efficiency

What it is An important characteristic of fan performance is efficiency. *Fan efficiency, simply defined, is the ratio of the rate of energy put into the air by the fan to the rate of energy put into the fan by the motor (shaft horsepower).* It is expressed as a percentage. (Typical efficiency curves are illustrated in Figures 2-11 and 2-18.)

Efficiency variations by fan type Different types of fans exhibit different efficiencies. In general, centrifugal fans (Figure 2-12) are the most efficient, with variations depending on blade design. Backward-curved airfoil blades are the most efficient of all, with forward-curved next, and radial blades the least efficient. Among the axials (Figures 2-13 through 2-15), propeller fans are the least efficient, tubeaxials next, and vaneaxials best. A combination type fan called tubular centrifugal (Figure 2-16) has close to centrifugal fan efficiencies but suffers some inefficiency because of the directional change of the air flow after it leaves the impeller proper.

Fan laws Fans exhibit some very predictable characteristics, and how these characteristics or parameters vary with one another is expressed in relationships called fan laws. The relationships are between fan air flow (expressed in cubic feet per minute), rotational speed (expressed in revolutions per minute), diameter, pressure, power required, and air density. Of most interest to operators are those parameters that can be measured or changed on-site and that affect the performance of a fan in a particular application. The following relationships, therefore, are of particular interest to an operator of an existing facility.

$$CFM \sim RPM$$
$$Pr \sim (RPM)^2$$
$$BHP \sim (CFM)^3$$

where:
   CFM = air flow in cubic feet per minute
   RPM = fan speed in revolutions per minute
   Pr   = total pressure in inches of water column
   BHP = fan shaft power or brake horsepower

(*Note:* ~ means "is proportional to" or "varies directly with.")

If the above relationships were to be expressed in words, they would look like this:

Air quantity (in cubic feet per minute) varies directly with speed (in revolutions per minute).
Pressure varies directly as the square of speed (in revolutions per minute).
Power varies directly as the cube of air quantity (in cubic feet per minute).

These are very important relationships, and their full meaning should be studied and understood.

Example applying fan laws   For instance, if you changed the speed (revolutions per minute) of a fan and, let's say, increased it by 10 percent, the air quantity (cubic feet per minute) should also increase by 10 percent, all other parameters (like pressure) remaining equal. The same modification with no change in air quantity would increase the pressure by 21 percent (110 percent × 110 percent = 121 percent). Most importantly, because this affects energy use, *if you were to increase air quantity (cubic feet per* *minute*) *by a factor of two (that is, double it), the power required would go up eight times* ($2 \times 2 \times 2 = 8$). *By the same token, if you were to cut the air quantity in half, the power requirement would be one-eighth of its original value.*

Fan noise Fan noise is an important criterion in proper selection and is frequently a problem in existing air-handling system installations. Fan noise is a function of design, volume flow rate (cubic feet per minute), total pressure, and efficiency. Normally, the quietest fans will be those operating in their most efficient range; therefore, efficiency has to be a *chief* criterion in picking fans. Low outlet velocity alone does not guarantee quiet operation, nor does low rotational or tip speed. Noise levels generated by fans are generally reported by manufacturers and supplied along with the other performance data.

Air Flow Adjustments and Fan Curves

Once a new fan system is installed, it is virtually always necessary to make some adjustments in the volume of air handled by the fan, and there may be other reasons later on to readjust the air flow. *The only way to affect the air flow through a fan is to change the characteristic either of the distribution system or of the fan.*

Discharge dampers If you decide to change the distribution system characteristic (that is, make the change external to the fan), this is accomplished with discharge dampers or other restrictions which reduce flow. (See the discussion in the section entitled "Distribution System and Ductwork" on page 70.) However, such a flow control technique also increases system pressure and therefore leads to increased power consumption by the fan. Closing down external dampers, however, is probably the least expensive of all flow control techniques.

Variable frequency (fan motor) controllers A better method of changing flow rate is by changing the fan curve. The most power-efficient way of doing this is by varying the rotational speed (revolutions per minute) of the fan. Indeed, some of the most efficient fan output controllers on variable air volume systems today are controllers that electronically vary a fan's speed (by varying the frequency of electricity driving the motor).

Inlet vanes Another technique that changes the fan curve characteristic is inlet vane control. *Inlet vanes* are dampers (sometimes called *vortex dampers*) which not only restrict but tend to spin the air as it enters the fan impeller. This is probably the most common method of fan output control, though it is not as efficient in reducing power consumption as the previous method (varying the revolutions per minute).

Variable pitch blades The output (cubic feet per minute) of tubeaxial and vaneaxial fans with radial blades can be controlled by adjusting pitch on the blades. This can be done as the fan operates. (Pitch is the angle the blades make as they cut through the air.) This technique generally retains reasonably high efficiencies over a wide range of conditions. In terms of noise generation, variable speed is preferable to variable pitch; but both techniques are considerably less noisy than inlet vane or outlet damper control.

DISTRIBUTION SYSTEM AND DUCTWORK

Definition and Purpose

The term *distribution system*—or the equivalent term *ductwork*—refers to virtually all the air passages external to the fan, from fan outlet to the occupied space, and from the space back to the fan's intake or the building's exhaust louver. The purpose of the distribution system, of course, is to provide a path for the proper quantities of air to travel from the fan to the spaces requiring it, and then to return that air from the spaces for reuse or exhaust. A distribution system might also be said to consist of housings, plenums, or shafts when these act as conduits used for the transport of air. See also the section entitled "Housings and Enclosures" which begins on page 70.

How Distribution System Affects Fan Performance

An air distribution system affects overall system performance because it imposes resistance to air flow. In general, the larger the air passages are, the less their resistance, but there are other design features that also have a big impact on system resistance. Some of these are turns, size transitions, aspect ratio (ratio of width to height), and damper type and placement. A fan's job is to drive an adequate amount of air through the distribution system and to draw it back; the fan selected must produce adequate pressure to do so. Thus, *the fan's capacity to deliver air must match the system's resistance to air flow.*

Distribution System Curve

Just as fans themselves have a characteristic curve (pressure versus air flow rate), so does a distribution system have one. Unlike a fan curve, however, this system curve is always in the same form: that of a parabola (curve of increasing steepness) emanating from the zero point of the pressure–air quantity graph. A typical system curve is shown in Figure 2-19. The basic relationship is:

$$Pr \sim CFM^2$$

where:

Pr = pressure needed to drive air through ductwork
CFM = system air flow in cubic feet per minute

(*Note:* ~ means "is proportional to" or "varies directly with.")

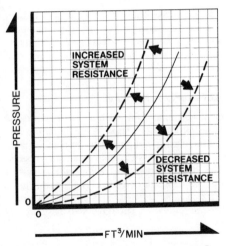

FIG. 2-19 A typical air distribution system curve: Parabolic in format, it expresses how much pressure is needed to force a given quantity of air through ductwork. The curve will pivot right or left as system resistance changes.

**Effect of Air Flow Adjustment
on System Curve**

The system curve can be changed by increasing or reducing resistance. Indicated also in Figure 2-19 is the effect of changing
resistance on the basic curve. Resistance in a system, of course,
is increased by closing down dampers or imposing other restrictions. If resistance is added to a system, the curve is moved to the
left (really, pivoted around the 0–0 axis). The opposite occurs if
system resistance is reduced (in that case, the curve is pivoted
clockwise).

The adjustment devices are called *balancing dampers* and are
located sometimes at the fan but always at various points in the
ductwork. Often registers and diffusers are equipped with integral
dampers. These are almost always manual, however, and once
adjusted, stay in their set position.

Relationship of System Curve to Fan Curve

A key point to remember is that *a fan operates on its fan curve
and a system performs on its system curve.* A fan pumps air into
a system, or to put it more exactly, it pumps air through a system
"circuit." The point at which the system and the fan *must* operate
is that point at which the two curves meet. *If the fan-plus-system
combination (that is, the air-handling system) is not operating at
the point you want it to, you must then take steps to change either
the system curve or the fan curve.* These methods have already
been discussed.

HOUSINGS AND ENCLOSURES

The term *housings and enclosures* refers to the surfaces (other
than the distribution ductwork) surrounding and/or connecting

components of an air-handling system. In the case of a prefabricated air-handling unit, the unit is shipped with integral enclosures and/or housings. See Figure 2-1. On larger systems, however, the fan housing is built in the field to suit the particular installation. Enclosures are also necessary to serve as outside-air and return-air mixing chambers and to provide a surrounding air chamber or plenum for filter sections, bypass sections, and coils.

BASES AND VIBRATION ISOLATION AND SOUND CONTROL

Overview

Most air-handling systems installed for comfort air-conditioning are not just plunked down on the floor of a mechanical room. Special provisions to accommodate the equipment are usually provided for reasons of (1) housekeeping, (2) vibration isolation, and (3) noise control.

First of all, a piece of air-handling equipment is usually placed on a base that is raised from the floor some 4 to 6 in. This base is called a housekeeping pad, and it simply provides a raised spot on which the unit can be set to facilitate clean-up of debris, water, and so forth in the surrounding mechanical room. Housekeeping pads do not have any appreciable vibration isolation or noise control function (except by virtue of their sheer mass).

Vibration and noise, though, can be very important criteria in equipment selection and other design choices—and they have occasionally been the source of some very knotty problems in existing installations. The two subjects are treated together here because they are very much related: Noise is simply vibration that manifests itself as unwanted sound. It should be realized that what follows is far from a comprehensive treatment of the subject of vibration and noise. The discussion here attempts only to hit some of the high points.

Basic Concept

A basic point to understand about vibration and sound is the source-path-receiver concept: Both sound and vibration are created by a *source*, transmitted along a *path*, and impinge upon a *receiver*. Reduction or isolation measures may be applied to any one of these elements.

Types of Vibration and Sound

When discussing comfort air-conditioning applications, sound and vibration are sometimes described in terms of their means of transmission. It can be structure-borne, or it can be airborne. Vibration and sound are further categorized by source. A common type of noise in air-handling systems is air noise, examples being air rushing past a grille or damper and noise caused by high air velocity in a duct. Vibration and sounds generated can also be high frequency or low frequency, and each requires different ways of solving the problems caused.

Resonance

One key thing to watch out for in dealing with vibrations from air-handling equipment particularly is resonant frequency; that is where the frequency of the vibrating source (for example, the revolutions per minute of a fan) matches or is close to the natural frequency of the mass of the assembly supporting the vibrating (rotating) element (for example, an air-handling unit suspended on a vibration isolation base). When the frequency of a vibrating source matches the natural frequency of an object free to vibrate, resonance occurs: The vibrational amplitudes (deflections) of the object are reinforced and severe, even violent, oscillations can occur. This is not something you want to have happen since equipment damage can result. It is possible to avoid this situation by careful examination of the frequency to be generated and the mass of the objects free to vibrate.

Steps to Avoid Vibration and Sound Problems

Some of the common steps taken to alleviate vibration and sound problems in air-handling systems are listed as follows.

Vibration isolators   These consist of springs and rubber or cork pads (or a combination thereof) supporting an assembly containing rotating equipment. Vibration isolators can be seen under the return fan in Figure 2-1.

Inertia bases   An *inertia base* is a thick, heavy, reinforced concrete mass on which the assembly containing vibrating machinery (a fan, for example) is mounted to give the entire assembly more mass for the purpose of changing its natural frequency.

Attenuation   *Attenuation* is the absorption, or reduction by other means, of sound transmitted through ductwork or surfaces. Examples of attenuation steps are the installation of built-up sandwich panel fan housings, duct sound lining, sound traps in ductwork, and high-mass walls or partitions. An in-place sound trap can be seen in Figure 2-1. Attenuation is also achieved across walls by plugging the gaps around pipes, ducts, and conduits passing through the walls.

Ductwork isolation   A rotating (vibrating) assembly must be completely isolated from the surrounding structure and ductwork, and for this purpose flexible connections are generally placed between the ductwork and fan—on both intake and discharge if applicable. These are simply unpainted canvas collars connecting the two adjacent air passages; the collars contain the air as it flows through but do not transmit vibrations.

MOTORS AND DRIVES

Types of Electric Motor Drives

Associated with every fan is a means of driving it; almost always, electric motors are used for this purpose. The means of connecting the shaft of the motor to that of the fan is called the *drive,* and fans can either be *direct-driven* (that is, coupled directly to the motor shaft) or *belt-driven* (that is, connected to the motor shaft by means of belts and pulleys). On larger fans, a belt drive is most common.

Adjusting Fan Speed

Virtually all motors used for fan drive purposes are constant speed—either 1140 rpm or, more commonly, 1760 rpm. Since either speed is generally faster than you would want most large fan rotors to turn, a belt drive can be used to allow the fan to turn at a somewhat slower speed. This is done by using different size pulleys (often called *sheaves*) on the motor and fan shafts.

Changing sheave diameter   To allow for adjustment of the speed of a fan, its motor is usually equipped with an adjustable sheave.

By adjusting the effective diameter of the sheave around which the belt makes contact, you can change the ratio of the diameters of motor sheave and fan sheave. The speeds of the shafts will be inversely proportional to the diameters of their sheaves. For example, a fan with a sheave four times the diameter of the motor sheave will rotate at one-quarter the motor speed—or 1760/4 = 440 rpm.

Changing belt tension   As the diameter of a motor sheave is adjusted, the distance between shafts for a given belt size changes somewhat also. Therefore, it is necessary that the motor be able to be moved slightly so that its distance from the fan shaft can be lengthened or shortened to maintain the proper amount of belt tension. Thus, motors are mounted on "rails" so they can slide back and forth.

Motor Power Requirements

For air-handling system applications, it is common to size a motor so that it will not overload with that particular fan at *any* operating condition. Because of this and since motors only come in nominal sizes, motors inevitably are oversized for the job they are called upon to do. However, since motors run at constant speed, the amount of power they require will drop if the load they drive is less than their rated full capacity load.

One speaks of motor horsepower, and one speaks of brake horsepower. *Motor horsepower* is the rated capacity of a motor to drive a load, while *brake horsepower* is the actual load imposed on the motor by whatever device it is driving—such as a fan.

Motor Efficiency

Motor efficiency relates the rate of energy input to the rate of energy output: The ratio of output to input is called motor efficiency and it will generally range from 80 to more than 90 percent. Most larger motors have an efficiency of 90 percent or better.

Motor Heat

Energy put into a motor goes one of two ways: into the fan shaft to rotate the fan (move and pressurize the air); or out as waste heat (inefficiency). If a motor is in a fan system's airstream, that waste heat goes into the airstream. If the motor is located outside the fan housing and not in the airstream, the heat of inefficiency goes into the ambient air. The shaft energy is generally considerably larger than the energy lost as heat.

Effect of Fan Energy on Airstream

When looking at larger fan systems, it is important to understand the effect that fan energy added to an airstream can have on airstream temperature. The airstream of a high-pressure, high-velocity fan system contains a lot of kinetic energy when it is moving at high speed, and thus would not exhibit much temperature rise from fan energy at that point. However, *as the airstream slows down* and reduces in pressure as it approaches and eventually exits into a space, essentially coming to rest, *the airstream's temperature does in fact rise. The rise is due to the energy added to the air by the fan motor.* A rule of thumb is that for every inch of pressure a fan system generates, almost 0.5 °F is added to the airstream's temperature.

Motor Control Equipment

Motors require motor control equipment; this equipment may have several functions. Motor control equipment may provide (1) a means of disconnecting motor and controller from the power supply; (2) a means of starting and stopping the motor; (3) protection from short circuiting and overheating; (4) protection of motor branch circuit conductors, control apparatus, and the operator; and (5) control of motor speed. The type of motor control equipment used depends on the size and type of motor, the power supply, and the degree of automation. Control equipment may be manual, semiautomatic, or fully automatic. A widely used type of motor controller for air-handling system applications is the across-the-line magnetic controller.

AUTOMATIC TEMPERATURE CONTROL

Purpose

Automatic temperature control might be thought of as the brains or the nervous system of the air-handling system. The purpose of automatic temperature controls is fourfold: (1) maintain conditions for human comfort, (2) maintain conditions of safety and security, (3) maximize capital equipment life, and (4) minimize energy use and cost.

Basics of Automatic Temperature Control

The following is an overview of a basic automatic temperature control scheme, which introduces some of the terms used. See

Figure 2-20, which depicts a simple but typical control situation for an air-handling system.

In a typical control situation, a *sensor* provides information about a *system parameter* (variable) to a *controller,* which actuates a *controlled device* (actuator) to obtain a desired *setpoint.* The *command* (output) of the controller must operate the controlled device in such a way as to maintain the setpoint. This must happen continuously and fast enough to maintain the setpoint—but not so fast that overshooting and hunting occur.

In the diagram, the coil discharge or *cold-deck temperature* (CDT) is the system parameter, and the *temperature sensor* sends *information* (input) about that temperature to the *CDT controller* (sometimes called a *receiver/controller*). The controller, in turn, sends a *command* (output) to the *cooling coil valve* (the controlled device), and the valve accordingly opens or closes to maintain the system parameter (by modulating the flow of chilled water through the coil).

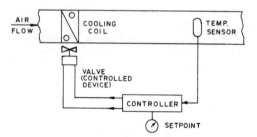

FIG. 2-20 A basic automatic temperature control scheme, applied to part of an air-handling system; the main elements involved are a sensor, a controller, and a controlled device or actuator. (Diagram courtesy Computer Controls.)

The input and output for this example are both *analog;* that is, a temperature can have any number of values, or, in other words, is infinitely variable. Likewise, the controller's output can command a setpoint of any value. In a different control scheme, if the input were the run status of a motor (on or off), the input would be *binary* (two-state). A command might also be two-state: start or stop.

What's to Be Controlled

The system parameters and types of equipment that are commonly controlled (through commands or output) on air-handling systems are listed here.

Fans and fan motors   These include supply, return, and exhaust fans which are started and stopped periodically. They may also be stopped for safety reasons (smoke, high temperature).

Outside-air quantity   This, along with return and exhaust air quantities, is varied by manipulating dampers.

Air temperatures   Common air-handling system temperatures controlled are listed below.

Preheated air   Resulting from how a preheat coil is controlled.

Mixed air   Resulting from the relative quantities of outside and return air that are blended together.

Cooled air (cold-deck)   Leaving the cooling coil; usually resulting from control of cooling coil valve or face-and-bypass dampers but sometimes from control of the chilled-water temperature itself.

Heated air (hot-deck)   Similarly resulting from control of heating coil valves or, again, face-and-bypass dampers. Can be controlled in broadbrush fashion by varying hot water temperature as well.

Supply air   May be one and the same as one of the above, but in any case it's the air delivered to the ductwork leaving the fan room.

Room air (space)   Control of room temperature is one of the ultimate objects of air-handling systems.

Humidity   Room humidity is what we really care about, but there may be high limits for airstreams to prevent condensation or other malfunctions.

Supply-air quantity   Applicable to variable air volume systems; methods generally used to control this were discussed in the section entitled "Air Flow Adjustments and Fan Curves" which begins on page 66.

Air cleanliness   Control here can vary from completely manual to highly automatic, depending on the sophistication of equipment used.

What's to Be Measured and Monitored

As opposed to the previous list, the following list describes specific items of information or input—that is, those values and states that need to be known in order to make informed control decisions. While the list is not necessarily complete, these are the most common system parameters that are measured by control systems.

Run status   Whether a piece of equipment, like a fan, is running or not; not important locally (because run status is usually obvious), but helpful where control is remote.

Filter cleanliness A filter is considered dirty when its resistance to air flow reaches a certain value, as measured by pressure drop across the filter.

Temperatures The most common are outside-air, return-air, mixed-air, cold-deck, hot-deck, supply-air, and room temperatures. Even a controlled temperature must be measured if it is to be controlled.

Flow An actual value of air flow is not commonly measured by control system instruments unless the air-handling system is of the variable volume type. More usual is simply an indication of whether there is flow in a system of not. This parameter is called *air flow prove*.

Damper position This is commonly determined by measuring the value of the pressure controlling the damper (if a pneumatic damper motor); whether a damper is closed or not can be indicated by a limit switch.

Current draw This is helpful to know on motors where input varies, such as motors for variable volume system fans.

Humidity The actual parameter measured automatically by instrument is either dewpoint or relative humidity read directly. Hand-held instruments can read wet-bulb temperature.

Devices Used to Control and Measure

Devices that physically control a particular parameter are called *actuators,* and devices that monitor the value or status of a parameter are called *sensors*. The most common of these used in air-handling systems are listed here.

Switch relay and starter   Used to start and stop electrical devices or monitor their status.

Damper motor   Can be pneumatic or electric.

Valve motor   Similar to damper motor.

Temperature sensors   Vary depending on the application. For duct temperatures, there are rigid airstream bulbs and flexible capillaries. Capillaries can be averaging type or low-limit type. Finally, space thermostats have built-in temperature-measuring elements that are different from all of these.

Electric current   If connected to the electric meter, this device may be a pulse counter. If a meter connection is not possible, current transformers attached to the main electric feeders are used.

Flow condition   Called flow "prove," the flow condition is useful to know—especially should flow stop unexpectedly. Sensors to detect flow in airstreams are either "sail" switches or pressure differential switches.

Control Strategies

What is a control strategy?   As applied to air-handling systems, a control strategy is a plan of action to be taken as the result of certain events occurring or of certain other data available. It involves the monitoring, receipt, and storage of information and the issuing of commands. It also involves a methodology or plan for formulating a specific course of action based on information received or known (stored).

Simple control strategy   In its very simplest form, a control strategy could involve a single control point, a manual means of control, and a decision by a person. An example would be starting a fan by hand at 8 A.M. In this case, the datum available is the time, the plan is the intended operating schedule of the fan in the mind of the operator, and the command is starting the fan (flipping a switch).

Moderately complex control strategies   Control strategies become more complex as automatic means replace human judgment, as variable input data are added to the stored data, and as multiple points of command and/or input are added.

For example, maintaining a leaving warm-air temperature at a fixed point by controlling a hot water valve involves some automation, a variable input (the measured leaving air temperature), and stored data (the desired air temperature). The actuator is the valve motor, the sensor is the leaving air temperature measuring device, the stored datum is the desired setpoint, and the "plan" is to keep the leaving air temperature at the setpoint by opening or closing the hot water valve. Although simple, this action has all the elements of a control strategy.

Complex control strategies   Control strategies get more complicated as other input or output points are added. For instance, if we wanted to vary the setpoint for the leaving air temperature according to how cold it is outside, we would introduce another measured input parameter—namely, outside-air temperature. The OA temperature measurement would become the basis for reset of the setpoint, and the measured leaving air temperature would have to be controlled to match that variable setpoint.

Control strategies get even more complex as other factors are introduced. Typical examples are electricity monitoring and demand control, enthalpy control, and start-time optimization. (These are common energy strategies offered by vendors of package energy management systems.) Demand control, for instance, requires instantaneous input on electricity usage and the time interval. There must also be access to such stored information as equipment currently operating, electric capacities of that equipment, and actions possible through the stopping and starting of certain equipment in certain prearranged sequences. The "off" time of a piece of equipment, moreover, may then be varied according to space temperature inputs. Another factor taken into account is what piece of equipment was last off and how recently. The plan that specifies how all this information is to be assembled and processed and what commands need to be sent out is a control strategy.

Note: As control strategies get more complex and replace human tasks, some operators may become concerned about their future role. While this concern is understandable, wise operators realize that controls are tools to help them do their jobs better and to relieve them of routine tasks. Their time is thus made freer, allowing them to be more effective in performing tasks that only human beings can do.

Control Strategy Hardware

Sensors and actuators   As is obvious from the previous discussion, there are first of all sensors and actuators, the sensors used to provide the measured input and the actuators used to receive and convey the output of a control strategy. A means of transmitting the data is necessary too, though this can be done in a fairly straightforward way via wires, including telephone wires. (The details of how signals are put in, transported, and taken out will not be discussed here.)

Strategy center   The other major hardware element for effecting control strategies is a strategy "center." In its simplest, nonautomatic form, this is the human operator. In its simplest automated form, it may be an assembly of relays, pneumatic-electric (PE) and electric-pneumatic (EP) switches, discriminators, receiver/controllers, etc. These are all common elements of a *conventional* pneumatic control system. The *characteristics of these devices and the way they are hooked up*—as they would be in a typical pneumatic system—*determine a control strategy*. With pneumatics, this is a "locked in" (that is, nonflexible) strategy.

Microprocessors   Electronic direct digital control (DDC) microprocessors dedicated to local control of equipment *provide more flexibility* than a pneumatic strategy center. *They are programmable, meaning they can accept a plan (a control strategy) via coded instructions*—instructions which can be easily varied at a later time to modify the strategy. A similar change in strategy with a setup involving relays, switches, etc. (that is, a pneumatic panel) would require changes in the types of devices used as well as changes in the way they were hooked up. Programmable microprocessors are thus much more flexible than pneumatic controls.

Central automation system   Another variant of the strategy center is the centralized automation system, sometimes called an *energy management system*. In most applications, these are only *partially devoted to control, much of their function being monitoring;* very often much of the localized control is left to the local control apparatus with its own built-in strategies.

3

Description of Air-Handling Systems

An air-handling system in a building consists of all the components described previously, including the distribution system and ductwork serving the spaces getting air. While the components in various air-handling systems may be similar, there are certain differences in the way these components are put together or in other characteristics that distinguish one system from another.

HOW AIR-HANDLING SYSTEMS ARE CATEGORIZED

Like fans, air-handling systems are categorized in several different ways.

Constant versus Variable Air Volume

First, air-handling systems can have constant air volume or variable air volume. (For *volume*, read *flow*.) Most system types are constant volume even though that term is not usually used to identify them. Variable volume systems, however, do usually carry the identifying tag *variable air volume* (VAV).

High-, Medium-, or Low-Pressure

Air-handling systems are also classified as high-pressure, medium-pressure, or low-pressure; these ratings correspond to the pressure ranges outlined for fans. The advantage of a high-pressure system is that more air can be forced through smaller ductwork, thus saving space and sheet metal cost; sometimes the benefit is just being able to get the air required to the rooms served in the space available. However, high-pressure systems tend to be energy inefficient: The fan power needed increases directly as pressure increases.

Single versus Double Duct

Another classification refers to the number of supply airstreams leaving the air-handling unit or fan room. Examples are single-duct systems and double-duct (or dual-duct) systems. The term *double-duct* refers to airstreams at different temperatures, one hot and the other cold.

All-Air versus Air-Water

Air-handling systems are also referred to as all-air or air-water systems. Of course, virtually all air-handling systems require the transfer of energy either into or out of the airstream, and, more times than not, this is done with hydronics ("wet system"). Therefore, few air-handling systems are truly "all-air."

This classification refers to the heating and cooling medium used *outside* the fan room. *If a space is heated or cooled entirely by the air being supplied to it, the system is an all-air system.* (An example is a double-duct system with mixing boxes.) *If, on the other hand, the system relies on both air and water outside the fan room, then it is referred to as an air-water system.* (An example would be a terminal reheat system where water is piped to the booster coils in the vicinity of the occupied spaces.) Also, any HVAC system which otherwise would be all-air—such as a double-duct-with-mixing-box system—is referred to as air-water if it is supplemented by a "wet" perimeter heat system, such as fin tube.

Central versus Local

Air-handling systems are also referred to as central or local—referring to where the system, but mostly the fan room, is located. A double-duct air-handling system with multiple mixing boxes scattered throughout a building is undoubtedly a central air-handling system. However, fan coil units located in the individual perimeter spaces are local air-handling systems.

Type of Zone Control

The main way air-handling systems are classified, however, is according to what is referred to as "type." These "type" names are the names in common use. Upon examination, what this classification uses to distinguish one air-handling system from another seems to be *the means and degree to which the system is capable of controlling the temperatures of the zones served.* This will become apparent as each system type is discussed.

AIR-HANDLING SYSTEM TYPES

The following sections describe the air-handling system types most likely to be found in commercial and institutional buildings and outline the key features and the pros and cons of each.

Single Zone (Heat-Cool-Off) (Figure 3-1)

Key features   This is a very common and probably the simplest type of air-handling system—more prevalent, however, in small sizes. Its main characteristic is that it only has a single zone of control, regardless of how many spaces it serves. Diagrams of a typical single-zone system are shown in Figure 3-1. Most domestic heating and air-conditioning systems are of this type, and so are most small self-contained packaged units (2 through 20 tons, say).

Note on the diagram the cooling coil and heating coil in the airstream. These are controlled in sequence so that simultaneous heating and cooling does not occur. When the thermostat requires cooling, the cooling coil operates and the heating coil is prevented from operating—and vice-versa when heating is required. With the proper thermostat, a *deadband* can also be provided (a deadband is a range of room temperature for which neither heating nor cooling is provided).

As can be seen, the room thermostat (which can only be placed in *one* of the spaces served) controls the coils directly and *in sequence*. On some small systems, the fan may also be linked to the control thermostat; in these systems, the fan only runs on a call for cooling or on a call for heat. (This setup is not feasible—or in conformance with most codes—if a continuous air supply is required during occupancy.)

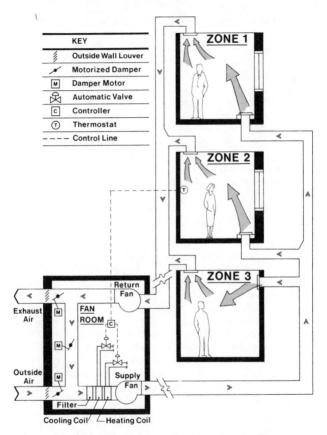

KEY

⫽	Outside Wall Louver
✎	Motorized Damper
Ⓜ	Damper Motor
⧖	Automatic Valve
©	Controller
Ⓣ	Thermostat
– – –	Control Line

ZONE 1

ZONE 2

ZONE 3

Return Fan

Exhaust Air

FAN ROOM

Outside Air

Supply Fan

Filter

Cooling Coil — Heating Coil

FIG. 3-1 Simplified diagram of a typical single-zone air-handling system. Note that each zone (or room) gets same temperature air, but only one zone controls it.

Advantages The advantages of this type of system are that it is simple, has only a single supply duct, is energy efficient (simultaneous heating and cooling is eliminated), and is inexpensive.

Disadvantages Its disadvantages are that there is no temperature control differentiation among spaces supplied and, mainly for that reason, the number and variety of spaces the system can suitably serve is limited. Further, a system that operates in this manner may not exhibit very good summertime humidity control since, when a space's dry-bulb temperature requirement is satisfied, dehumidification (by the cooling coil) stops also.

Single-Duct, Terminal Reheat (Figure 3-2)

Key features Figure 3-2 diagrams a typical system of this type. Note the similarity in configuration to the single-zone system. *The difference is that in the terminal reheat system control zones are provided wherever desired by inserting reheat coils in the branch ducts to those zones*. Each reheat coil is controlled by a zone thermostat located in the space served (or a typical room in that space).

The term *reheat* implies how this system works: A low temperature is supplied from the fan to the distribution system year round. In winter, this temperature is obtained by mixing appropriate amounts of cold outside air and return air; in summer, it is obtained by using the cooling coil. This air supply, commonly in the range of 55 to 60 °F, is then reheated by each reheat coil as needed to meet each space's temperature requirement.

Reheat coils are generally supplied with hot water of about 140 °F, although lower-temperature water can be used if the coils are designed for it. For large systems where higher pressures are necessary to distribute air throughout the system, the reheat coil may be integral with a constant volume pressure reduction box,

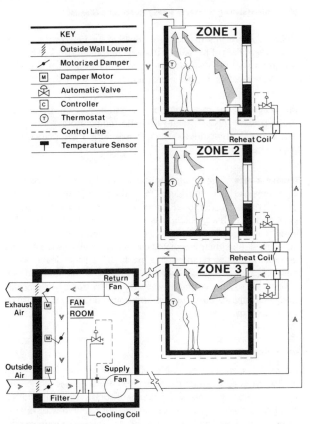

KEY

⧄	Outside Wall Louver
⬎	Motorized Damper
M	Damper Motor
⊠	Automatic Valve
C	Controller
T	Thermostat
---	Control Line
⊤	Temperature Sensor

ZONE 1

ZONE 2

ZONE 3

Reheat Coil

Reheat Coil

Return Fan

Exhaust Air

FAN ROOM

Outside Air

Supply Fan

Filter

Cooling Coil

FIG. 3-2 Typical single-duct, terminal reheat system (three zones shown, but there could be many). Each space has thermostat, which controls temperature of air to each space. This system affords best control, but uses much energy.

which automatically maintains a constant air flow to the spaces served regardless of pressure variations in the ductwork. This feature also makes the system more flexible (makes it easier to modify control zones after original construction).

Advantages   The advantages of this system type are that it provides excellent humidity and dry-bulb temperature control and that, being a single-duct system, it has fairly conservative duct space requirements.

Disadvantages   The terminal reheat system has several disadvantages, however. It is considered to be the most energy inefficient system type of all (although there are certain steps that can be taken to make this problem less severe). It requires remote hydronic coils, valves, and pipes above the ceiling throughout the area served, plus their associated control lines; all this means that the system's space requirements are high. If water leaks occur, they can cause additional problems. While the air-handling portion proper is not expensive, the remotely located hydronic booster coils, associated piping, valves, and thermostats add on to make this a fairly expensive system.

Air-Mixing

Air-mixing systems are those in which the air supply is divided into two airstreams—one hot, one cold—for purposes of meeting widely varying space needs. Examples of such systems are the so-called multizone system and the double-duct system with mixing boxes. Figures 3-3 and 3-4 show these systems diagramatically, and their differences and similarities can be observed. You will note that in both cases zones are served by the same coils: one heating coil and one cooling coil. The hot and the cold airstreams

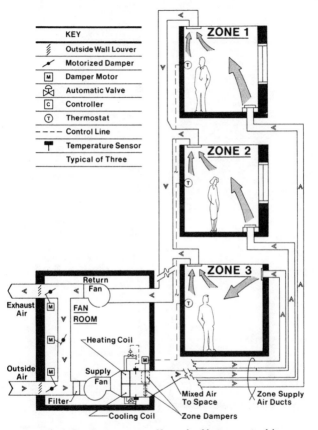

FIG. 3-3 Typical multizone system. Hot and cold air are mixed by zone dampers after leaving the coils, and each zone gets individually controlled air temperature; number of zones may be limited by space needed to run zone ducts.

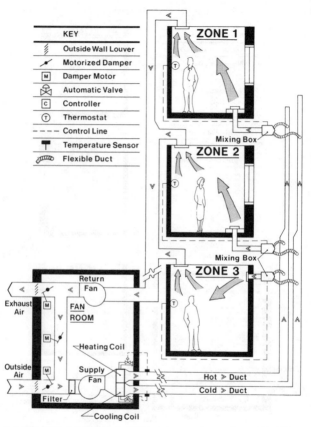

FIG. 3-4 Typical double-duct system with mixing boxes. Compare to multizone: Mixing occurs locally rather than in fan room. It can accommodate many zones, each of which requires a mixing box.

from these coils are then mixed to give the supply-air temperature required by the zone being served (according to the zone thermostat). The main way that multizone and double-duct systems differ, from the control standpoint, is in where the mixing is done.

Air-Mixing: Multizone (Figure 3-3)

Key features *Multizone units* are generally factory built, preassembled units complete with fan, heating coil, cooling coil, zone mixing dampers, and sometimes a mixed-air plenum on the intake. As seen in the diagram, mixing occurs immediately after the air passes through the cold and hot decks and is controlled by a set of zone mixing dampers for each zone. Zone mixing dampers respond to the zone thermostat, located in a typical room of the zone served.

Advantages The advantages of the multizone system are that it is modest in cost (provided there are not too many zones or that duct runs don't get too long), that it affords individual zone control, and that the major components can be purchased prefabricated.

Disadvantages A major disadvantage is that such systems are limited in terms of the size and number of zones they can serve. The space required for ducts can be considerable, especially close to the fan room. If there is a large number of zones, the sheet metal cost can become prohibitive. The system is not too good on humidity control, and it is not flexible (that is, it is difficult to add or modify zones).

Air-Mixing: Double-Duct with Mixing Boxes (Figure 3-4)

Key features   In this system, the means for mixing the hot and cold airstreams is not built into the unit but must be installed separately as mixing boxes. These are installed remotely, one near each zone being served, like the reheat coils on the single-duct terminal reheat system. Mixing boxes, however, do not require any hydronic connections since this is an all-air system.

Double-duct systems can serve large areas within a building— sometimes as much as fifteen to twenty floors of a high rise, for instance. For large systems, it's usually necessary to go to a high-pressure system to keep duct sizes reasonable. The mixing box then acts as a pressure reduction and noise reduction device as well as a constant volume and mixing device.

With the double-duct system, both a cold duct and a hot duct are installed throughout the building, wherever there are areas to be served.

Advantages   The advantages of the double-duct system with mixing boxes is that it is simple to understand, provides reasonably good control of temperature (although not of humidity), and does not entail any hydronic systems beyond the fan room.

Disadvantages   The disadvantages are that it takes a considerable amount of space (especially since there are two ducts), is expensive, is not especially energy efficient, and has some humidity control problems. (There are certain steps that can be taken, however, to improve its energy performance, such as discharge-air reset.) See O&M Procedure 4, page 13, and ECM 2, page 109.

Induction Unit (Figure 3-5)

Key features   In a sense an induction unit is a mixing system too, but in this case locally induced room air is mixed with centrally treated primary air. The mixing takes place locally, in the unit, which is either in the occupied space itself (under-window unit) or in the ceiling plenum above (horizontal ceiling unit). The induction unit is a specially engineered device which "induces" recirculated room air to flow across a coil into the unit. The coil contained in the induction unit can be for both heating and cooling (controlled by a two-position valve), or the unit can have two sets of coils.

Although the primary air is conditioned centrally, the major energy transfer of an induction unit system occurs at the coil locally. The coil is usually hydronic and must be supplied with both hot water and chilled water.

One centrally located, high-pressure, primary air fan serves numerous induction units throughout a building. Because the ratio of locally induced air to centrally distributed primary air is about 3 or 4:1, the amount and size of distribution ductwork is much smaller than for other systems, where all circulated air comes from a central point.

The primary air temperature is usually controlled on a preset schedule related to the outside air condition, while the local room wall thermostat (or built-in thermostat) controls the coil. Each induction unit is also equipped with a filter, and the cooling coil is equipped with a drip pan (in case of condensation), which must be piped to drain.

Advantages   One advantage of an induction unit system is low first cost, particularly when applied to large, multiple-unit instal-

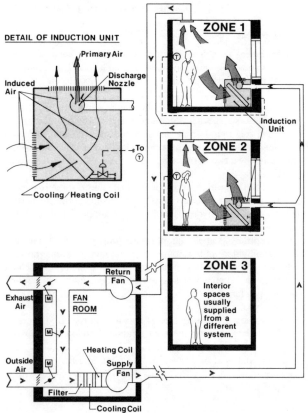

FIG. 3-5 Typical induction unit system. Primary air discharging from nozzle creates low pressure, *induces* air from space to flow across cooling or heating coil, which handles *most* of space load. Because primary air quantity is low and at high pressure, ducts can be small.

lations where the reduced size of ductwork makes a difference. Another plus is individual room control. As mentioned before, duct sizes are minimized, which makes this type of system advantageous for buildings with little ceiling or shaft space.

Disadvantages The disadvantages are that generally it serves perimeter spaces only, although it is possible to get horizontal ceiling-hung units for interior areas. More commonly, however, a different system type entirely is used for the interior. A further disadvantage is that the induction unit itself requires maintenance *in* the occupied space; this can include filter changing, valve maintenance, coil cleaning, etc. Should condensation occur on the cooling coil, a wet coil surface and drip pan—together with dirt accumulated in a filter—make it difficult to maintain.

Variable Air Volume (Figure 3-6)

Key features All air-handling systems talked about so far are constant air volume systems in which zone temperature control is achieved by varying the air supply temperature. *The variable air volume system achieves control by varying the quantity of air supplied while maintaining a more or less constant supply temperature.*

In general, the basic variable air volume (VAV) system is designed to handle only cooling loads, and relies on a space's internal heat (lights, people) or supplementary heat for any heating requirements. Supplementary heat may be perimeter baseboard or fin tube radiation; a perimeter constant volume, variable temperature air system; or even a small reheat coil in the supply ductwork.

A VAV system is most effective where space loads vary a lot during the course of a typical day. An example might be a building with

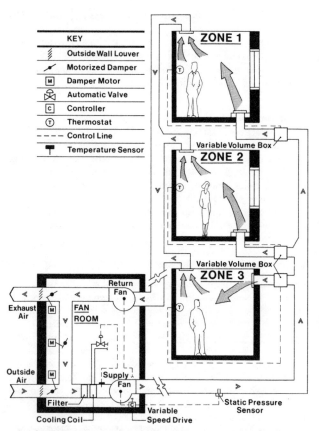

FIG. 3-6 Typical variable air volume (VAV) system. Supply temperature is constant and cool, and zone loads are met by varying air through zone-variable volume boxes; fan slows down to handle less air to match boxes. Most energy efficient system, mostly due to reduced fan motor energy.

multiple exposures of glass where, depending on sun position, the load on the various orientations changes: east in the morning, south at noon, and west in the afternoon. A constant volume system would have to be designed to supply air to meet the sum of the maximum loads of each orientation continuously. A VAV system, on the other hand, need be designed only to handle the maximum *instantaneous* load of all orientations. The air quantity then varies during the day and in effect "follows the load around."

Savings in operation of VAV systems (versus constant volume) result primarily from savings in fan motor power at reduced air volumes. Ideally, fan power is proportional to the cube of air quantity handled (see the section entitled "Fan Laws" which begins on page 64), so reductions in air flow have the potential of being triply effective in power reduction.

Air quantities on VAV fans can be controlled by various methods. Inlet vane control is common, but the characteristics are better with variable frequency controllers, though these are more expensive.

The variable air volume system is in essence a single-duct system, similar in appearance and layout to a single-zone reheat system. In lieu of reheat coils, however, there are variable volume devices located locally, associated with and controlled by thermostats in the zones served. Air entering a space at a constant low temperature (nominally 55 to 60 °F) has the potential to cool that space. Control is achieved by reducing the air's cooling capacity (reducing its flow), thus controlling to the desired space temperature.

Advantages   Variable air volume systems are relatively inexpensive, flexible (layout changes are relatively easy), provide individual space control, and, most important, are less expensive to operate than most other systems.

Disadvantages One problem with VAV systems is that they may overcool during periods of unoccupancy. This is because a VAV box may be set for a certain minimum airflow below which it cannot go. Even this minimum amount of air may be too much to meet the load when no lights are on and no people are present (that is, when there is no internal load). The space will then overcool, and it may take a while after a space is reoccupied for it to heat up again. Other disadvantages are poor humidity control and complaints of stuffiness at low airflow.

Another potential problem is poor air distribution patterns within a space at low air flow. If conventional diffusers and registers (which are selected for a specific air quantity) are used to discharge the air, the air may tend to "dump" or pour into a room at lower flows.

Fan Coil

Key features A *fan coil* air-handling *system* is really a series of small, local air circulation systems. A fan coil unit is a prepackaged device consisting of a small fan (with motor), a filter, and one or two coils. The unit may be a floor-mounted under-window type with finished cabinet that sits at the perimeter of a space, or it may be a horizontal ceiling-mounted unit with or without connecting ductwork. The principle in either case is the same.

Generally, the units operate on recirculation only. If ventilation air is required, it is usually necessary to have a separate means of supplying outside air (like windows that open) or an independent ducted system.

Advantages Advantages of a fan coil system are that it is inexpensive and provides individual space control; it eliminates the need for a central air-handling unit and so saves the space that unit would consume elsewhere.

Disadvantages Disadvantages include the need for local maintenance (including filter changing) in the occupied space, the blockage of the perimeter wall space taken up by the unit, and sometimes high noise levels.

Rooftop

Key features The term *rooftop* has come into common use, but it refers not so much to a type of system as to the unit's location. Rooftop units and the air distribution systems they serve may in fact be of various types, including single-zone, multizone, terminal reheat, or even variable volume.

Rooftop units are generally used in industrial applications, where low cost is of considerable importance, or as add-ons to condition a remodeled space. Large new commercial and institutional buildings are not usually designed around rooftop equipment.

Advantages The main advantages are that rooftop units are pre-packaged (including cooling, if desired), factory built, and inexpensive. Being located on the roof (or elsewhere outside the building envelope proper), they eliminate the need for space for comparable air-handling and air-conditioning equipment inside (though ductwork is still required).

Disadvantages The unit's being outside makes maintenance somewhat more difficult, especially in bad weather. Although rooftop units can be of good quality, there tends to be a large variation in the quality of equipment among vendors, and maintenance contracts from the suppliers are not always available.

Unitary

The term *unitary,* as applied to air-handling systems, refers to those systems that are made up of a number of small packaged units, each with an integral refrigeration cycle and air-handling capability. Examples are window air-conditioners, through-the-wall room air conditioners, air-to-air heat pumps, and water source heat pumps. These units may range in capacity all the way from fractional tonnage to 20 or 25 tons. Such multiple, packaged, unitary systems can be applied to almost all classes of buildings in certain situations. They are especially suitable when performance requirements are less demanding and when a simplified, inexpensive installation is desired. Because there are so many varieties and applications of unitary systems and because so many different types of equipment are involved, this manual will not attempt to explore the characteristics of such systems, nor will any further generalizations about their pros and cons be presented.

4

Moderate Cost Energy Conservation Measures (ECMs) to Improve Energy Efficiency

The O&M procedures described in Chapter 1 are intended for implementation by in-house staff, and the discussion of them is directed not only toward helping staff detect *when* those procedures would be applicable but also to telling them *how* to go about implementing them. The discussion of energy conservation measures (ECMs) that follows here, however, is mostly directed toward understanding *when* the ECMs might be applicable. Because they're more costly and complex than the O&M procedures, most would require outside contractual help to implement. In addition, outside consulting services might be required to prepare specifications and other documents for use in soliciting contractors' bids.

RECOMMENDED ECMs

ECM 1: Programmed Start/Stop

What it is The rationale for turning equipment off during periods of unoccupancy or very low use was already discussed. (See O&M

Procedure 1, page 2.) While manual starting and stopping of equipment is simple, inexpensive, and uncomplicated, it can become burdensome and perhaps even too time consuming in buildings with lots of equipment. In such a situation, automatic means are more effective in assuring that the units are indeed started and stopped when required—and automatic start/stop frees up workers' time for more challenging and important tasks.

Means of implementing

Timers One method of automatically starting and stopping equipment is with 7-day timers. These are mechanical devices that contain a rotating disk on which "start" and "stop" pins can be placed to correspond to the desired "on" and "off" times. Each day of the week can have a different schedule. Generally, one timer is required for each piece of equipment.

Such devices should be periodically checked to see that they remain on time. When the number of timers in use becomes too large, this means of programmed start/stop becomes burdensome (for instance, following a power outage or a daylight savings time change, when they would all have to be reset).

Microprocessors Another effective means of starting and stopping equipment is the small microprocessor panel. Panels that control several loads are now available. Microprocessor panels cost more than mechanical timers but, if used to control several devices, can be almost as cost effective. It should also be recognized that, while some microprocessors are limited in functional capacity to starting and stopping equipment, some are also available with the capability of handling other functions as well.

Central automation system Programmed start/stop can also be accomplished with a centrally located automation system, or *energy management system* (*EMS*), as they are sometimes called. One should not install such a system, however, solely to perform start/ stop functions; that should be only one of many functions performed.

ECM 2: Automatic Mixed-Air, Cold-Deck, and Supply-Air Temperature Reset

What it is This is simply the similarly named manual procedure (see O&M Procedure 4, page 13) done by automatic means. The rationale and benefits were outlined in the description of O&M Procedure 4. It is important to be able to determine when it might be worthwhile to reset by automatic means, however.

Automatic reset means first that the process of resetting the air temperature takes place continuously, not just once per shift or whenever an operator can get to it. Second, the guesswork is taken out of the process: If the zone sampling is done correctly, the air temperature shouldn't ever be reset too high, causing space discomfort problems.

The principle of this ECM is to reset the mixed-air temperature (MAT) or cold-deck temperature (CDT) as high as possible without jeopardizing comfort conditions in any space. Therefore, some or all space conditions must be monitored, with the "worst case" situation being the basis for resetting the temperature.

When it's applicable The potential for savings from this measure is greatest for large fan systems and where there is evidence that the supply-air temperature can be raised a substantial amount above its nominal setting without causing space overheating or humidity problems.

For instance, if the nominal, or "worst case," setting on your MAT is 55 °F, but you know from experience that a good deal of the time the spaces will all cool adequately with an air temperature of 63 °F, that is evidence that considerable savings could result from air temperature reset. If you know, however, that there is always some space that requires 55 °F air (or within a few degrees of that) regardless of occupancy conditions, weather, sun position, etc., then there would be little benefit in implementing this ECM.

Sampling space conditions   The best way to sample space conditions is to monitor whether the space thermostats are satisfied. This can be done pneumatically by tapping into the control air pressures to the space thermostats and running those pressures through a high-pressure discriminator (selector), with the highest pressure passing through. This "pass-through" pressure is the reset signal to the MAT controller. As long as the selected or "worst case" pressure is in the "satisfied" range, the airstream temperature can be reset upwards. If it goes into the "unsatisfied" range, the air temperature needs to be reset downwards.

This ECM is most easily accomplished on multizone air-handling systems because in these systems all control signals are led back to the vicinity of the fan itself. It is a simple matter to tie into the control signal lines and connect them as described above. On air-handling systems where the thermostatic control signals are *not* led back to the fan room (such as double-duct-with-mixing-box systems or terminal reheat systems), there may be many scattered control zones. Thus, it may not be practical to sample all zones (too costly). In this situation, it is necessary to be selective and pick a representative few.

The basis for selection of zones is the "worst case": Which zone, under certain conditions, is likely to be calling for the coldest air? For example, you might pick a conference room zone (where, with a meeting in progress, maximum cooling would be needed), a west-facing zone (maximum afternoon sunload), representative zones from other exposures (like south and east), and any other zone where overheating is known to occur. You want to pick sample zones such that, if they are *all* kept satisfied, the chance of any other space in the building overheating is minimal.

How CDT reset differs from MAT reset   CDT reset is similar to MAT reset, except that it applies to the air temperature downstream of the cooling coil during the cooling season. Since the cooling coil is generally also the only means of dehumidification, you must be careful not to reset the temperature so high that space humidity gets too high (even though the dry-bulb temperature is satisfactory). It is thus necessary to monitor relative humidity in the spaces (return airstream is OK) and use that parameter as a limit on CDT reset: Should the relative humidity (RH) reach, say, 55 percent, no more upward reset is permitted to take place, and if RH exceeds 55 percent, downward reset must take place.

Interface with existing controller   When installing an automatic temperature reset system, one requires a *receiver/controller* that can accept a reset signal. (A receiver/controller is the device that controls an air temperature to a setpoint.) If your present receiver/controllers can't, they would have to be adapted or replaced. This adds considerably to the cost of implementing this ECM. (However, where substantial reset can occur, the payback is still quite favorable.)

Monitoring results   After this ECM is put in place, an effort should be made to monitor the results by keeping track of what the actual MATs or CDTs are and whether the spaces being served remain satisfied. *The best indication of success would be that these temperatures are running substantially (say, 5 to 10 °F) above their nominal low setpoint a good deal of the time and no spaces are* *experiencing comfort problems.* If this is the case, you can rest assured that considerable energy is being saved.

Typical annual savings (based on Chicago area weather and utility rates) would run about $150 per degree of reset per 1000 ft³/min for MAT reset, about $80 for CDT reset. Payback ranges are 2 to 4 years and 3 to 6 years respectively. (These figures are based on mid-1985 costs.)

ECM 3: Outside Air Quantity Adjustment

What it is   The technique of taking measurements of temperature to determine actual percentage of outside air was discussed in Chapter 1. See O&M Procedure 2, page 5. That procedure described adjustment of OA quantity to the design value. This ECM entails taking a closer look at the required minimum OA quantity and determining whether the design quantity is excessive.

Why it's important   Based on personal experience in examining many existing hospital systems, the author can say that reassessing the amount of outside air required and comparing it to what is actually being supplied is definitely worthwhile. Since, during colder weather, systems that operate with recirculated air frequently operate at minimum OA quantities, savings in the amount of OA that needs to be heated (or, in hot weather, cooled) can be significant. And since the capital involved in making the adjustment is minimal, payback can be very fast (under 1 year is not uncommon).

When it's applicable   In order to determine whether such adjustments should be made, you need to collect some data and perhaps make some measurements.

A format for assembling this information is shown as Table 4-1. Let's look more closely at the data needed to complete this table.

1. It is first necessary to know for each air-handling system the quantities of air involved. This includes not only total supply air but also total exhaust air from the area served. These data could come from a reliable test and balance report; otherwise, you will have to take new measurements.

2. Next, and very important, it is necessary to determine the amount of outside air now being supplied at the minimum setting. The technique for doing this was explained in O&M Procedure 2, page 5.

3. The next tasks involve gathering and tabulating data that will help you determine what the minimum OA quantity should be. First, the average maximum number of people expected to be present in the area served by each air-handling system should be estimated and tabulated. This should not be the sum of the maximum number of people that each space *could* hold, but rather the total average maximum that realistically might be expected over a 1 to 2 hour interval.

4. You then want to list the OA quantities required by the governing code or regulations for your particular facility. This may be your local building code, the HEW (Department of Health, Education, and Welfare) requirements for hospitals (Table A-1 in Appendix), or state requirements. (You may need several columns for these data.)

5. You then may want to list the OA quantity recommended by good engineering practice. In this regard, the ASHRAE requirements might be a good source. These could be obtained

TABLE 4-1 Summary of Outside Air Quantities

FAN SYSTEM	TOTAL AIR QTY.	EXHAUST AIR QTY.	MAX PEOPLE	OUTSIDE AIR QUANTITY					PERCENT OF TOTAL
				EXISTING	BASIS	CODE	ASHRAE	RECOMM	
(tag #)	(ft³/min)	(ft³/min)	(average)	(ft³/min)	(AC*/hr)	(ft³/min)	(ft³/min)	(ft³/min)	(%)

Totals									

*AC = air changes.

from the latest ASHRAE Handbook[1] or the latest edition of ASHRAE Standard 62-1981 on ventilation[2].

6. Once you have tabulated all these data, you are then ready to examine them and make a determination on whether your actual minimum OA settings can be reduced. In places where this can be done, you should list the recommended quantity in a separate column. Then you are ready to implement the changes.

Implementation The steps outlined in O&M Procedure 3, page 10 should be followed.

ECM 4: Supply Air Quantity Adjustment

What it is This ECM involves reducing the total air supply of a fan system when there is evidence it is excessive.

When it's applicable This ECM is most likely to apply when there have been modifications to the code requirements or there have been space load changes. Examples of things which might cause space load changes are lighting reductions, reflective film or windows, new window glass, solar load reduction due to new construction, and equipment removal.

Typical signs of excess supply air are a space that tends to overcool or a supply airstream that has a relatively high temperature even when the space it serves requires full cooling.

If excess supply air is suspected, an outside engineering con-

---

[1] *ASHRAE Handbook and Product Directory: Applications Volume,* American Society of Heating, Refrigerating and Air-Conditioning Engineers, Inc., Atlanta, Georgia, 1981.

[2] "ASHRAE Standard 62-1981, Ventilation for Acceptable Indoor Air Quality," American Society of Heating, Refrigerating and Air-Conditioning Engineers, Inc., Atlanta, Georgia, 1981.

sultant should be engaged to make a detailed survey and determination of the extent of reduction possible. A form for summarizing the data *on a fan system basis* is shown as Table 4-2. As with the outside air quantity evaluation, this form can be used for a summary of *total system* data, but is not designed for a room-by-room evaluation.

Implementation If an evaluation is made and the results indicate that a reduction can be made, it is necessary then to rebalance the entire system so that the air is still properly distributed. This should only be attempted in-house if you have sufficient personnel and expertise to accomplish an entire system rebalance. Otherwise, you may need to engage an outside test and balance contractor to do it for you.

In any case, if substantial reductions of supply air quantity can indeed be made, the savings in labor hours and annual energy requirements can be considerable. (See the discussion in the section entitled "Fan Laws" which begins on page 64. Note particularly the law relating power to air quantity (in cubic feet per minute), and the "Example Applying Fan Laws" on page 65.)

ECM 5: Two-Speed Fan Operation

What it is This ECM is closely related to ECM 1, Programmed Start/Stop, in that it represents something of a middle ground between that ECM and doing nothing. This ECM assumes that, while it would not be possible to eliminate the air supply entirely to the areas being served at any time, it would indeed be possible to cut it down sometimes—by about half, say. As with ECM 4, the power–air quantity relationship governs, and a reduction by half in the number of cubic feet per minute supplied during reduced-

TABLE 4-2 Summary of Total Air System Supply Quantities

FAN SYSTEM	APPROX. AREA SERVED	TOTAL AIR QUANTITY			
		EXISTING	CODE BASIS	CODE	RECOMM
(tag #)	(ft²)	(ft³/min)	(AC*/hr)	(ft³/min)	(ft³/min)

Totals										

*AC = air changes.

occupancy periods would yield a power consumption of only about one-eighth of that during full operation.

When it's applicable Two-speed fan operation is applicable when there are periods during which an area served has much less intensive use or markedly reduced activity or occupancy—yet when there is enough going on there to prohibit the complete shutting down of the air-handling system.

Again, a survey of certain operating data and equipment capacities is required. A sample table on which to tabulate this information is shown as Table 4-3. For any systems where it is suspected that this ECM would apply, the data should be assembled and examined.

Note that one key indication of a change in activity level is a big drop in the number of people in a given area during certain shifts. (Ask department heads for this information.) If there is a considerable drop in the number of people and the area is not critical, then a reduction in the amount of air supplied during the reduced-occupancy period should be considered. Once the information has been collected and tabulated, however, the final determination should be made only by key personnel. They should include the director of engineering (or person in the equivalent position) and the heads of the departments affected.

Implementation Outside assistance for both engineering specifications and installation are required for this ECM.

ECM 6: Economizer Control and Heating Coil Sequencing

What it is Economizer control and heating coil sequencing is intended to correct a control sequence sometimes designed into

TABLE 4-3 Proposed Reduced-Capacity Fan Operation

SUPPLY FAN TAG #	MOTOR HP	RETURN OR EXHAUST FAN TAG #	MOTOR HP	PEOPLE BY SHIFT (A-B-C)	RECOMMENDED TURNDOWN	AREA SERVED

single- or two-zone air-handling systems that causes unnecessary waste. (The control sequence is one which would be necessary, though, were the system to have many zones.)

The sequence that should be looked for on one- or two-zone systems is this: (1) An economizer control methodology is being used to maintain a constant or fixed mixed-air temperature (MAT), and (2) a space thermostat controls a downstream heating coil to meet space heating needs. The problem is that the economizer control, for a good part of the time, is keeping the MAT lower than it needs to be (by bringing in more outside air than necessary).

The proposed modification to this control sequence is to have the controlling thermostat—in most cases, the room thermostat—sequence the economizer control with the heating coil control such that the air supplied is first warmed by closing the outside air dampers to minimum position *before* the heating coil is allowed to come on. This makes it possible to take maximum advantage of the heat in the return air before using any "new energy" (that is, the heating coil) to meet space heating needs.

How to spot it   These situations can usually be easily spotted by looking at the control drawings for an air-handling system. In the field, they can be spotted by noting whether the mixed-air temperature on a system is being set and maintained at a constant fixed temperature regardless of whether one or more spaces served are calling for heat.

Implementation   Although the control system modification is fairly simple, an outside engineer or control contractor should be engaged to verify that this ECM is applicable, and a control contractor should implement it.

ECM 7: Morning Warm-up and Cool-down Sequence

What it is   A morning warm-up and cool-down sequence involves the starting and stopping of certain equipment and might be thought of as related to, and an enhancement of, ECM 1, Programmed Start/Stop.

It is customary, in large buildings where some air-handling systems are shut down during periods of unoccupancy, for those systems to be started at some interval prior to expected occupancy in order to bring the space temperature up (or down) to an acceptable level by the time of occupancy. (Space temperature may have been set back—or allowed to run up—during the unoccupied period.) In other words, a fan system might be started 1 or 2 hours prior to an expected occupancy time of 8 A.M. This is commonly done by an operator who goes around and flips the switches on the fans on a prearranged schedule, but it may also be done on a preprogrammed basis by a timer.

In either case, *more often than not the equipment is run during those hours just prior to occupancy with the outside air (OA) dampers open to their minimum (or greater) positions*. This is because the OA dampers were designed to open whenever the fans came on. Since generally the only reason for bringing in OA is for human air-freshening purposes, there is no need to provide that ventilation prior to occupancy. During cold or hot weather the OA that is brought in must be either cooled or heated, and thus is consuming energy unnecessarily. (In this case, "cold" is less than 50 °F, "hot" more than 70 °F.)

In hot, dry climates, however, where high ambient humidity is not the norm, it may be advantageous to bring in *full* OA during the period before occupancy if that OA is less than, say, 70 °F. This would be a *cool-down sequence,* since the "free" OA would then be used to "purge" a building that had had a heat buildup overnight.

When it's applicable   It is easy to tell when this ECM is applicable: Simply observe the OA damper position of the supply fans when they run in the morning prior to occupancy. Do the OA dampers remain closed with the recirculating dampers full open? If not, this ECM is a good candidate. (*Note:* This ECM is *not* applicable to fans that run continuously.)

Implementation   The way to implement this ECM is to install a means of keeping the OA dampers shut until occupancy occurs, regardless of whether the fan is running. Physically, what this entails is keeping the damper motor control pressure (on pneumatic systems) from opening the OA dampers. One way to do this is simply to install an electropneumatic (EP) switch, which prevents the air pressure from getting through, and then put control of that EP switch on a timed basis. If the fan motor itself is controlled from a 7-day timer, a second 7-day timer can control the EP switch to the dampers. This second timer would be set to the occupancy hours while the fan motor timer would be set to the desired hours of fan system operation.

A similar result could be accomplished through microprocessor control or through a central automation system since both the fan motor signal and the damper EP switch are binary outputs. Of course, if the fan motors are controlled manually, the damper motor EP switch can also be controlled manually, but manual control would entail an additional trip to the fan room every day to open the dampers.

5

Instrumentation

Instrumentation, as discussed here, includes all devices used for measurement and monitoring of the parameters relating to an air-handling system's operation and control. Periodic checks and proper maintenance of controls and data-recording instrumentation are essential to assure reliable and economical system operation.

TYPES OF DEVICES

A variety of air handler controls and instruments are now in use. Such instrumentation generally employs one of five types of control media (or some combination thereof): mechanical, hydraulic, pneumatic, chemical, and electronic. Each type has advantages and disadvantages, and each requires proper maintenance to ensure reliable functioning. Some common problems with the various types of devices are listed as follows.

1. Mechanical devices may suffer from wear, binding, or misalignment, which impairs repeatability of settings.
2. Hydraulic and pneumatic devices are subject to leaks and plugging of sensing and control lines.
3. Chemically activated devices are subject to leakage and contamination.
4. Electronic devices may malfunction because of circuit problems, dirty or loose contacts, or insulation failure. These devices may also be damaged by exposure to heat or excessive vibration.

In each case, the manufacturer's specific guidelines on maintenance should be consulted. Past records of control system malfunctions and the frequency of "drift" in the accuracy of an instrument are good guides to a reasonable maintenance schedule for inspection, readjustment, and recalibration.

DATA SAMPLING TECHNIQUES

The location where a data sample is taken is very important when measuring temperatures or pressures. It can be as important as the selection of the proper measuring device.

One should avoid taking readings immediately downstream of dampers, bends, or fans. Airstreams in such areas can stratify and/or form pockets, leading to errors, especially when readings are taken at a single point in the duct cross section.

When a single-point probe (such as a thermocouple) is used, readings should be made at several points in the duct (especially if the duct is a large one) in order to get a representative average.

Unless representative data are obtained, the test results will be of little value.

TEMPERATURE MEASUREMENT

A liquid-in-glass or bimetallic thermometer may be used for temperature readings where accuracy is not critical. It should be calibrated by measuring a known standard (such as boiling water or ice water) and then adjusting your readings accordingly. For quick, more accurate readings, a thermocouple device may be used. Table 5-1 gives more information on these devices.

HUMIDITY MEASUREMENT

Instruments used for measuring the humidity of air are called *hygrometers*. A *psychrometer* is a particular kind of hygrometer consisting of two temperature sensors, one partly covered with a cloth wick or sock. The wick is wetted with distilled water and ventilated with air moving at a sufficient rate (generally 700 ft/min or more) across the instrument. The resultant evaporative cooling produces an approximate wet-bulb temperature.

Sling Psychrometer

In a sling psychrometer, the two thermometers are mounted side by side in a frame fitted with a handle for whirling the device through the air. See Figure 5-1.

Charts or tables are available from ASHRAE and some equipment manufacturers showing the relationship between temperatures and humidity; these charts can be used to determine relative

TABLE 5-1 Applicability of Various Temperature-Measuring Devices

Device	Application	Range (°F)	Precision (F°)	Limitations
Liquid-in-glass thermometer	Gases and liquids by contact	Mercury 38 to 575 Alcohol 0 to 100	Less than ±0.1 to 10	In gases, accuracy affected by radiation
Bimetallic thermometer	Approximate readings	0 to 1000	Nominally ± 1, usually more margin of error	Time lag; unreliable
Thermocouple (copper-constantan)	Rapid readings	Up to 700	±0.1 to 15	Subject to oxidation

FIG. 5-1 Sling psychrometer. (Photo courtesy Sybron/Taylor.)

humidity. (The ASHRAE psychrometric chart is shown in Figure 1-2 on pages 20 and 21.)

When air temperatures are below 32 °F, water on the wick may either freeze or supercool. A psychrometer can be used at high temperatures, but care must be taken to be sure that the wick remains wet.

Direct Reading Gauges

Many organic materials change dimension with changes in humidity. This property has been utilized in the design of a number of simple and effective humidity indicators.

Organic materials commonly used are human hair, nylon, dacron, animal membrane, wood, and paper. Unfortunately, no organic material has yet been found that will consistently reproduce this action over an extended period of time, and responses can be significantly affected by exposure to extremes of humidity. These devices require initial calibration and frequent recalibration or setting. Nevertheless, they are useful because they can be arranged to read directly in terms of relative humidity, and are simpler and less expensive than most other types of humidity-measuring devices.

Types of instruments for reading humidity are listed in Table 5-2, with brief comments on each.

VELOCITY MEASUREMENT

The flow of air is usually measured at or near atmospheric pressure. Under this condition, the air can be treated as an incompressible fluid, and simplified formulas can be used with sufficient precision. A device used for measuring the velocity of a fluid is commonly called an *anemometer*. There are several types of anemometers.

Deflecting-Vane Anemometer

The *deflecting-vane anemometer* consists of a pivoted vane enclosed in a casing. See Figure 5-2. Air exerts pressure on the vane as it passes through the instrument from an upstream to a downstream opening. The vane movement is resisted by a spring and a dampening magnet. The instrument gives instantaneous readings of directional velocities on an indicating scale. When the velocity is fluctuating, it is necessary to visually average the needle swings to obtain an average velocity. This instrument is useful for studying air motion in a room and for locating objectionable drafts.

TABLE 5-2 Applicability of Various Humidity-Measuring Devices

Device	Application	Range (°F)	Precision (% RH)	Limitations
Psychrometer	Room air, outside air, moving air in ducts	0 to 500	±0.3 to 3	Should be used in airstreams moving about 1000 ft/min; difficult to use at subfreezing temperatures
Dimensional change; mechanical	Very slowly moving air	−40 to 150	±3	Frequent calibration required; adversely affected by temperature above 125 °F and RH below 20%
Electrical conductivity	Measurement and control	−40 to 150	±1.5 to 3	Susceptible to damage by air contaminants; frequent calibration required

FIG. 5-2 Deflecting-vane anemometer. (Photo courtesy Davis Instrument.)

Revolving-Vane Anemometer

The *propeller,* or *revolving-vane, anemometer* consists of a light, revolving wheel connected through a gear train to a set of dials. Each instrument requires individual calibration. At low velocities the friction drag of the mechanism is considerable.

Pitot Tube

A *pitot tube,* used in conjunction with a manometer, provides a simple method of determining air velocity at a point in a flow field. The equation for determining air velocity from measured velocity pressure is:

$$V = 1096.5 \sqrt{\frac{h_w}{\rho}}$$

where:

V = velocity in feet per minute

h_w = velocity pressure (from pitot tube and manometer reading) in inches of water

ρ = density of air in pounds per cubic foot

Since velocity in a duct is seldom uniform across any section, and since a pitot tube reading indicates velocity at only one location, a traverse of the duct is usually made to determine the average velocity. In general, velocity is lowest near the edges or corners of the duct, and greatest at or near the center. Suggested pitot tube locations for traversing round and rectangular ducts are shown in Figure 5-3.

Since low peripheral velocities are not measured, air flow calculated by this method is usually slightly higher than the actual. Traverses at duct inlets and outlets require special techniques. Pulsating or disturbed flow will lead to errors in measurements. If possible, the pitot tube readings should be taken at least 7.5 diameters downstream from a disturbance such as a turn.

Hot-Wire Anemometer

Measurement of low air velocities (0 to 100 ft/min) is difficult. Often the flow pattern is unstable. If a sensing element is heated

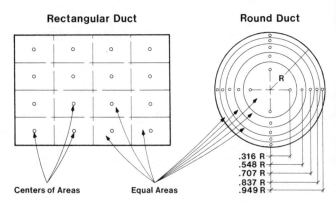

FIG. 5-3 Recommended pitot tube traverse pattern for rectangular and round ducts.

electrically at a fixed rate and exposed to an airstream, the temperature difference between the element and the airstream becomes a measure of velocity. In a *hot-wire anemometer,* a very fine heated wire is used as a resistance thermometer element whose temperature may be determined accurately.

The main advantages of a hot-wire anemometer are (1) suitability for transient velocity and turbulence measurements, (2) good accuracy in measuring low velocities, and (3) availability of specialized sensors and accessories. The correct calibration of hot-wire anemometers requires consideration of the effects of temperature, humidity, and pressure on air properties.

Instruments for measurement of air velocity are described further in Table 5-3.

TABLE 5-3 Characteristics of Instruments for Measuring Air Velocity

Device	Application	Range (ft/min)	Precision	Limitations
Deflecting-vane anemometer	Used to measure air velocities in rooms, outlets	30 to 24,000	±5%	Not well suited for duct readings; needs periodic calibration
Revolving-vane anemometer	Used to measure moderate air velocities in rooms, ducts	100 to 3,000	±5 to 20%	Extremely subject to error; easily damaged; needs periodic calibration
Pitot tube	Is the standard instrument for measuring duct velocities	180 to 10,000	±1 to 5%	Accuracy falls off at low end of range
Hot-wire anemometer	Used to measure low or transient velocities and turbulent flows	1 to 1,000	±1 to 20%	Requires accurate calibration at frequent intervals; complex; costly

PRESSURE MEASUREMENT

The pressure of the air in the sheet metal ducts ordinarily used in air-handling systems is very small—seldom more than 0.3 lb/in² and more often less than half of this. Nevertheless, these pressures have a large influence upon performance of the fan and delivery of air through the various parts of the duct system.

Static Pressure

Any fluid, including air, will exert pressure on the walls of the container (duct) in which it is confined. Thus, a gauge connected to a tank of compressed air will indicate what is called static pressure: the "outward push" of the air against the walls of the tank.

In a duct with no air movement (say, because a damper has been shut tight), the static pressure of the air can be measured by means of a gauge connected to a tube inserted through the wall of the duct as shown in Figure 5-4. In this same figure, a hinged vane is shown in the airstream. If the air were at rest, the vane would hang vertically. However, if the damper at the outlet of the duct were opened and the air started to move, the vane would be tipped back by the impact of the moving air against its face. The pressure on the face of the vane which holds it at an angle would be due entirely to the velocity of the moving stream of air. If the vane were hung parallel to the flow of air, its angle would not be affected by the movement of air. The pressure on the two faces would be equal because static pressure only is exerted against them. Similarly, even when air is in motion, a tube with its opening perpendicular to the direction of air flow, as shown in Figure 5-4, will not be affected by the movement of the air. A gauge connected to the tube will indicate only the static pressure, or sideward push, of the air, exactly as it would were the air at rest.

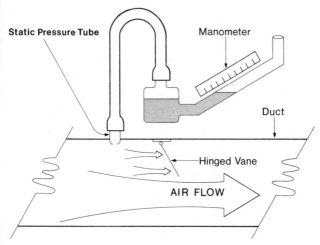

FIG. 5-4 Reading duct static pressure; pressure of air movement that deflects vane is *not* sensed by tube.

Total Pressure

On the other hand, if a tube is inserted into the duct with its opening facing into the moving airstream as in Figure 5-5, the pressure indicated by the gauge will be higher than when the static tube was used.

The reason for this is apparent when the cause for the tilt of the vane is considered. The vane swings up at an angle because of the pressure exerted upon its face by the moving stream of air. Similarly, a gauge connected to a tube with its open end facing the stream of air would indicate not only the static pressure but, in

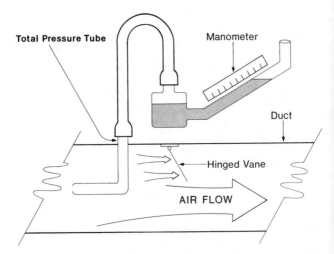

FIG. 5-5 Reading duct total pressure; tube senses *both* static pressure *and* added pressure caused by air movement.

addition, some pressure due to the movement of the stream of air. The greater the velocity, the greater the pressure exerted against the tube opening (or the face of the vane). The greater the velocity, the more the vane will be tilted. Similarly, the greater the velocity, the higher the pressure indicated by the gauge.

Velocity Pressure

This action of a moving stream of air offers a means of measuring the velocity of that air. The pressure measured in the way just described is known as total pressure. It is greater than static pressure because of the added pressure due to the movement of the airstream. If measurements are made of total and static pressure using two separate tubes, the difference in the measurements of the two gauges would give that excess pressure due only to the velocity: That excess pressure is known as velocity pressure.

To recap, there are three pressures:

1. Static pressure
2. Velocity pressure
3. Total pressure

Total pressure is the sum of static and velocity pressure. However, only total and static pressure can be measured; the velocity pressure is found by taking the difference.

Manometer

In general, for the purposes discussed here, we are primarily interested in *differential pressure:* the difference in pressure between two points in an airstream. The principal device used to measure differential pressure is the *manometer*. This device is so universally used that it is often the standard for calibration of other instruments. The manometer consists of a U-shaped glass tube partially filled with a liquid. The difference in the height of the two fluid columns is proportional to the pressure differential to be measured. This device is illustrated in Figure 5-6.

Manometer tubes should be kept chemically clean. The bore (inside diameter) is not very important. Bores of at least 0.1875 in. for rough, and 0.5 in. for more precise measurements are recommended. Liquids other than water are sometimes used for low-pressure measurement. When this is done, readings must be corrected for density of the fluid.

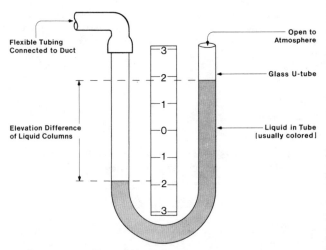

FIG. 5-6 U-tube mamometer; pressure in tube causes difference in elevations of liquid columns.

Draft Gauge

U-gauges used for measuring pressure differences of a few inches of water or less are often set at an angle for scale amplification. These are commonly called *draft gauges* or *inclined manometers*. In gauges of this type, only one tube of small bore is used and the other leg is replaced by a reservoir. The accuracy of the gauge depends on the slope of the tube, so the gauge base must be leveled carefully. A slope of less than 1 in 10 should not be used. Where pressures are read under extreme variations in temperature, it is necessary to correct for the change in density of the liquid in the manometer. An inclined manometer is illustrated in Figure 5-7.

ROTATIVE SPEED MEASUREMENT

Rotative speed is usually defined as the number of complete turns or revolutions an object turns per minute (rpm).

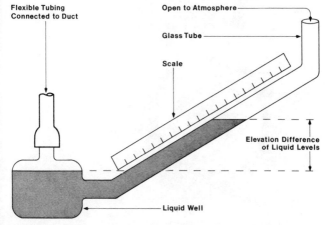

FIG. 5-7 Inclined manometer; used for reading small pressure differentials.

Tachometer

Tachometers are direct-measuring revolutions-per-minute counters used in checking the rotative speeds of motor and fan shafts. See Figure 5-8.

Stroboscopes

Optical revolutions-per-minute counters work by flashing a light at high speeds at a rotating object, increasing the number of flashes

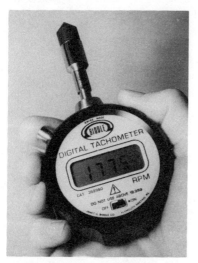

FIG. 5-8   Digital tachometer for measuring rotative speed. (Photo courtesy of Biddle Instruments.)

FIG. 5-9 Stroboscope; uses timed flashes of light to measure rotational speed. (Photo courtesy of Biddle Instruments.)

per minute until the optical effect of stopping the object is achieved. When the object appears stopped, the rotative speed is equal to the flashes per minute. See Figure 5-9.

ELECTRIC POWER MEASUREMENT

It is common, when measuring electric power, to measure the amperage and voltage in a circuit and then derive the power con-

TABLE 5-4 Measurement Techniques and Formulas for Determining Electric Power Consumption

Type of circuit and equation	Diagram	Typical voltage (V)
Single-phase, 2-wire: $kW = \dfrac{V \times A}{1000} \times PF$		115 or 230
Single-phase, 3-wire: $kW = \dfrac{(V_1 \times A_1) + (V_2 \times A_2)}{1000} \times PF$		115 (both) (230 between remaining two lines)

Three-phase, 3-wire:

$$kW = \frac{1.73 \times V \times A \times PF}{1000}$$

where: $A = \dfrac{A_1 + A_2 + A_3}{3}$

(Note: V can be measured between any two phases.)

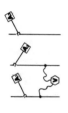

208, 230, or 463

Three-phase, 4-wire:

$$kW = \frac{1.73 \times V \times A \times PF}{1000}$$

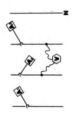

208 or 460 (with corresponding phase-to-neutral voltages of 115 and 277)

sumption using some simple formulas. A simplified procedure for this is shown in Table 5-4,

where:
- A = Current (amperes)
- V = Voltage (volts)
- PF = Power factor (in the absence of better data, use a power factor of 0.8 for motor loads; for lighting and electric heat, use 1.0)
- N = Neutral line (ground)
- ⊸ Ammeter (Amprobe) (see illustration, Figure 5-10)
- Ⓥ⌇ Voltmeter

FIG. 5-10 AMPROBE®, a clamp-on current-measuring device. (Photo courtesy of Amprobe Instrument.)

COSTS

Table 5-5 lists approximate costs (in 1985 dollars) of the instrumentation discussed here. These figures are provided here simply to give the reader a "ballpark" idea of the relative cost of different devices available. Specific quotations should be obtained before committing to any purchases.

TABLE 5-5 Cost Ranges of Some Air-Handling System Instrumentation

Measurement device	Cost range	Some manufacturers
Temperature: Liquid-in-glass thermometer	$6–$300	Cooper, Dwyer, Davis Stortz, Taylor
Bimetallic thermometer	$10–$26	Cooper, Stortz, Taylor, Weston
Thermocouple and digital readout	$150–$400	Beckman, Davis, Leeds & Northrop
Humidity: Sling psychrometer	$43–$80	Airserco, Bacharach, Davis, Foxboro, Princo, Stortz, Taylor
Digital readout	$375–$600	Bacharach, Beckman, Davis
Air velocity: Pitot tube	$80–$115	Alnor, Dieterich, Meriam, Mid-west Instruments
Thermal (hot-wire) anemometer	$400–$1175	Airguide Instruments, Alnor, Climet, Datametrics, Davis, Dwyer, Kurz, TSI

Instrument	Price range	Manufacturers
Vane anemometer	$400–$735	Airguide Instruments, Alnor, Climet, Datametrics, Davis, Dwyer, Kurz, Taylor, TSI
Air flow: Capture hood	$1500–$2000	Alnor, Brooks Datametrics, Short-ridge, TSI
Air pressure: Manometer	$100–$500	Bacharach, Datametrics, Davis, Dwyer, Meriam
Rotative speed: Tachometer	$175–$275	Biddle, Davis, Epic, Weston
Stroboscope	$175–$575	Biddle, Davis, Epic, Weston
Electric power: Ammeter	$92–$240	Amprobe, Biddle, Esterline-Angus, Weston
Voltmeter	$35–$240	Amprobe, Biddle, Esterline-Angus, Weston

Appendix

TABLE A-1 Department of Health and Human Services Requirements for Hospital Ventilation for Areas Affecting Patient Care*

Area designation	Air movement— relationship to adjacent area**	Min. air changes outside air per hour	Minimum total air changes per hour	Recirculated by means of room units	All air exhausted directly outdoors	Relative humidity (%)	Design temperature (°F)
Operating room	Out	4	20	No	—	45–60	70/75
Trauma room	Out	3	15	No	—	45–60	70/75
Delivery and birthing rooms	Out	3	15	No	—	45–60	70/75
X-ray, card. cath., spec. procedure	Out	3	15	No	—	45–60	70/75
Nursery suite	Out	1	6	No	—	30–60	75
Recovery room	—	—	6	No	—	30–60	75
Intensive care	—	—	6	No	—	30–60	70/75
Patient room	—	—	2	—	—	—	75
Patient corridor	—	—	2	—	—	—	—
Isolation room	In	—	6	No	Yes	—	70/75
Isolation alcove or anteroom	Out	—	10	No	Yes	—	—
Examination room	—	—	6	—	—	—	75
Medication room	—	—	4	—	—	—	—
Pharmacy	—	—	4	—	—	—	—
Treatment room	—	—	6	—	—	—	75
X-ray	—	—	6	—	—	—	75
Physical therapy and hydrotherapy	—	—	6	—	—	—	75
Soiled utility	In	—	10	No	Yes	—	—
Clean utility	—	—	4	—	—	—	—

Area	Air movement relationship**	Min. air changes of outdoor air per hr	Min. total air changes per hr	All air exhausted directly to outdoors	Recirculated within room units	Relative humidity / temp
Autopsy	(In)	—	12	Yes	No	—
Darkroom	(In)	—	10	Yes	—	—
Nonrefrigerated body holding room	(In)	—	10	Yes	—	—
Toilet room	(In)	—	10	Yes	—	70
Bedpan room	(In)	—	10	Yes	—	—
Bathroom	(In)	—	10	Yes	No	75
Janitors' closet	In	—	10	Yes	—	—
Sterilizer room (equipment)	(In)	—	10	Yes	No	—
Linen and trash chute rooms	In	—	10	Yes	—	—
Laboratory general	—	—	6	—	No	—
Biochemistry	(Out)	—	6	—	No	—
Histology	(In)	—	6	Yes	No	—
Bacteriology	(In)	—	6	Yes	No	—
Serology	(Out)	—	6	—	No	—
Glasswashing	(In)	—	10	Yes	—	—
Sterilizing	(In)	—	10	Yes	No	—
Food preparation center	(Out)	—	10	—	No	—
Warewashing	(In)	—	10	Yes	No	—
Dietary day storage	(In)	—	2	—	—	—
Laundry, general	—	—	10	Yes	No	—
Soiled linen (sorting and storage)	(In)	—	10	Yes	No	—
Clean linen	—	—	2	—	—	—
Anesthesia storage (see code requirements)	—	—	8	Yes	—	—
Central medical and surg. supply	—	—		—	—	(max) 70
Soiled room	(In)	—	6	Yes	No	—
Clean workroom	(Out)	—	4	—	No	75

*From DHHS Publication No. (HRS-M-HF) 84-1, July 1984.
**Parentheses denote less stringent air movement requirement relative to adjacent area.

TABLE A-2 Conversion Factors

Physical quantity	Conversion factor
Area	1 ft² = 0.0929 m²
Cooling capacity	1 ton = 3.52 kW
Density	1 lb/ft³ = 16.0 kg/m3
Enthalpy	1 Btu/lb = 2.33 kJ/kg
Heat flow rate	1 Btu/h = 0.293 W
Length	1 in. = 2.54 cm
	1 ft = 0.305 m
Power	1 hp = 0.746 kW
Pressure	1 in. WG = 249 Pa (N/m²)
	1 lb/in.² = 6895 Pa (N/m²)
Specific volume	1 ft³/lb = 0.0625 m³/kg
Temperature	T(°F) + [T(°C) (9/5) + 32]
Velocity	1 ft/min = 0.00508 m/s
Volume	1 in.³ = 16.4 mL
	1 ft³ = 0.0283 m³ or 28.3 L
Volume flow rate	1 ft³/min = 0.472 L/s

Bibliography

Air-Conditioning and Refrigeration Institute (ARI): *Refrigeration and Air-Conditioning,* Prentice-Hall, Inc., Englewood Cliffs, New Jersey, 1979.

Ambrose, E. R.: "Excessive Infiltration and Ventilation Air," *Heating, Piping and Air-Conditioning,* November, 1975.

————: "Upgrading System Performance in Existing Installations," *Heating, Piping and Air-Conditioning,* March 1976.

American Society of Heating, Refrigerating and Air-Conditioning Engineers, Inc.: *ASHRAE Handbook and Product Directory: Applications Volume,* Atlanta, Georgia, 1982.

————: *ASHRAE Handbook and Product Directory: Equipment Volume,* Atlanta, Georgia, 1983.

————: *ASHRAE Handbook and Product Directory: Systems Volume,* Atlanta, Georgia, 1984.

————: "ASHRAE Standard 90A-1980: Energy Conservation in New Building Design," Atlanta, Georgia, 1980.

————: "ASHRAE Standard 55-1981, Thermal Environmental Conditions for Human Occupancy" (ANSI B193.1-1976), Atlanta, Georgia, 1981.

————: "ASHRAE Standard 62-1981, Ventilation for Acceptable Indoor Air Quality," Atlanta, Georgia, 1981.

Bell, Morton: "Reducing the Cooling Load on Air Handling Systems," *Specifying Engineer,* January, 1977.

Bridgers, Frank H.: "Hospital Codes and Standards Versus Energy Conservation Design and Retrofit," Albuquerque, New Mexico, 1981.

Carrier Air-Conditioning Company: *Handbook of Air-Conditioning System Design,* McGraw-Hill, New York, 1965.

Clancy, Lena, and Robert Cowell: "Energy Cost in Food Delivery Calculated," *Hospitals,* February, 1980.

Dressler, F., and W. Spiegel: *Practical Energy Management in Health Care Institutions,* Blue Cross of Greater Philadelphia, Philadelphia, Pennsylvania, 1977.

Dubin-Mindell-Bloome Associates: "Guidelines for Saving Energy In Existing Buildings—Engineers, Architects and Operator's Manual—ECM 2," Federal Energy Administration, Catalog FE1.22:21, June, 1975.

"Energy Management: Commitment to Cost Containment," *Hospital Engineering,* January–February, 1979.

Gough, M.: "Hospital Energy Management Information System—Overview: Programs, Projects and Activities," The Hospital Research and Educational Trust, Chicago, Illinois, 1980.

Hirst, Eric: "Analysis of Energy Audits at 48 Hospitals," Oak Ridge National Laboratory, ORNL/CON-77, Oak Ridge, Tennessee, 1981.

Hirst, Eric, and James Stelson: "Studies in Energy Management: Institutional," *Heating, Piping and Air-Conditioning,* September, 1981.

Hittle, Douglas C., William H. Dolan, Donald J. Leverenz, and Richard Rundus: "Theory Meets Practice in a Full-Scale Heating, Ventilating and Air-Conditioning Laboratory," *ASHRAE Journal,* November, 1982.

Hospital Research and Education Trust: *Hospital Energy Management Procedures Workbook,* HRET Catalog 9350, Chicago, Illinois, 1979.

Kendrick, Lee: "Energy Conservation and Management for Buildings," Falls Church, Virginia, 1977.

Koehler, Walter: "Reduced Energy Use Occurs Despite Facility's Growth," *Specifying Engineer,* March, 1980.

Liptak, Bela G.: "Savings Through CO₂-Based Ventilation," *ASHRAE Journal,* July, 1079.

"M/E Update—Air Handling Units," *Specifying Engineer,* August, 1981.

National Fire Protection Association: *NFPA Standard 90A,* Air Conditioning Systems, 1985. Boston, Massachusetts.

Piper, James: "Air Handling Units—Preventive Management," *Building Operations Management,* 1981.

Reynolds, Smith, and Hills: *Energy Conservation Studies of Veterans Administration Hospitals,* V594P54, February, 1974.

Roy F. Weston, Inc.,: "Energy Conservation Guidelines Manual for HVAC Systems," Illinois Institute of Natural Resources (INR) Report ILLDOE-79/14, Springfield, Illinois, March, 1980.

———: "Energy Management—Engineering of Conservation Systems," American Institute of Plant Engineers (AIPE) Seminar, 1975.

Schultz, F. C.: "Converting Constant Volume Systems to VAV," *Specifying Engineer,* April, 1977.

———: "HVAC Adjustments Are Needed to Save Energy," *Specifying Engineer,* March, 1980.

Stroch, Hans H.: "Development of a Hospital Energy Management Index," *ASHRAE Transactions,* vol. 84, pt. 2, New York, 1978.

Strock, Clifford, and Richard L. Koral (Eds.): *Handbook of Air-Conditioning, Heating and Ventilating,* Industrial Press, Inc., New York, 1965.

The Trane Company: "Energy Conserving System Modifications," *Trane Engineer's Newsletter,* vol. 5, no. 8, November, 1976.

———: "Fans and Their Application in Air-Conditioning," LaCrosse, Wisconsin, 1971.

——— "Introduction to Control Application," LaCrosse, Wisconsin, 1970.

——— : "The State of the Art in Multizone Control," *Trane Engineer's Newsletter,* vol. 5, no. 4., May–June, 1976.

———: *Trane Air-Conditioning Manual,* LaCrosse, Wisconsin, 1965.

U.S. Department of Energy: *Building Energy Use Compilation and Analysis, Part C—Conservation Progress in Commercial Buildings* (draft), Buildings Div., Washington, D.C. May 1981.

———: "Energy Audit Workbook for Hospitals," Doc. DOE/CS/0041/3, Washington, D.C., September, 1978.

————: "Energy Measures and Energy Audits Grant Programs for Schools and Hospitals and for Buildings Owned by Units of Local Government and Public Care Institutionns," *Federal Register*, vol. 44, no. 64, April 2, 1979.

————: "Grant Programs for Schools and Hospitals and for Buildings Owned by Units of Local Government and Public Care Institutions, Amendment of Regulations," *Federal Register*, vol. 46, no. 98, May 21, 1981.

U.S. Department of Health, Education, and Welfare: "Energy Strategies for Health Care Institutions," United States Government Printing Office, HEW Publication (HRA) 76-620, Washington, D.C., April, 1976.

————: Enviro-Management and Research, Inc.: "Total Energy Management for Hospitals," HEW Publication (HRA) 77-613, Washington, D.C., 1977.

U.S. Department of Health and Human Services: *Minimum Requirements for Construction and Equipment of Hospital and Medical Facilities* (draft), Washington, D.C., 1982.

Uklesbay, Nan, and Kenneth Uklesbay: "Food Service Equipment and Energy Costs," *Hospitals*, March 16, 1980.

Index

ABOUT THE AUTHOR

David L. Grumman founded Grumman/Butkus
Associates (formerly Enercon, Ltd.), Evanston,
Illinois, in 1973. It provides energy consulting
and design engineering services to such clients
as Mercy Hospital, Northwestern University,
Amoco Corporation, and A. C. Nielsen Company. Prior to 1973, Mr. Grumman spent thirteen
years on the engineering staff of a nationally
known architectural/engineering firm in Chicago. He is a registered professional engineer in
ten states. He received his B.M.E. from Cornell
University in 1957 and served two years active
duty in the U.S. Navy's mine force. Mr. Grumman is active in the American Society of Heating, Refrigerating and Air Conditioning Engineers, was chairman of its Standards
Committee, and now serves on its Research &
Technical Committee, Environmental Health
Committee, and Technical Committee 4.8 on
Energy Resources. He has been widely pub-
lished in professional journals.